AF531592

INDUSTRIAL ENZYMOLOGY

ENCYCLOPAEDIA
OF
ENZYMOLOGY

Vol. III
INDUSTRIAL ENZYMOLOGY

By

Dr. Arvind N. Shukla

School of Studies of Zoology & Biotechnology

Vikram University

Ujjain

DISCOVERY PUBLISHING HOUSE PVT. LTD.

NEW DELHI-110 002

First Published-2008

ISBN 978-81-8356-357-4 (Set)

© Author

Published by:

DISCOVERY PUBLISHING HOUSE PVT. LTD.

4831/24, Ansari Road, Prahlad Street,
Darya Ganj, New Delhi-110002 (India)
Phone: 23279245 • Fax: 91-11-23253475
E-mail: dphbooks@rediffmail.com
dphtemp@indiatimes.com
Website: www.discoverypublishinghouse.com

Printed at:

Sachin Printers, Delhi

Preface

This classic *Encyclopaedia of Enzymology* has been compiled to provide an understanding of enzymology for university degree study and teaching. It comprehensively covers theory and applications of these biological life supporting catalysts in a readable book that will serve as a bridge to more advanced and specialized areas. It not only introduces the fundamentals and biochemistry of enzymes but also provides the molecular and experimental background. This unique piece of work compiled five volumes containing wealth of information for researchers, microbiologists, and biotechnologists. This excellent title is intended mainly for students taking degree course which have a substantial biochemistry component. The chief aim of this particular book is to help student understand the concepts involved in enzymology. An attempt has been made to give a perspective of each topic, and examples are quoted where appropriate.

There can be no claim to originality except in the manner of treatment and much of the information has been obtained from the books and scientific journals available in the different libraries.

The author express his thanks to his friends and colleagues whose continue inspirations have initiated him to bring out this book.

The author is painfully aware of the shortcomings, errors and misprints that have crept in, and shall be grateful to receive suggestion for improvement of the next edition from all the readers.

The author express his gratitude to Mr. Wasan and staff of M/s Discovery Publishing House Pvt. Ltd. for their whole hearted co-operation in the publication of this book.

Author

Contents

1

INTRODUCTION

Technical development, especially the dramatically rapid development in instrumentation during the past 30 years, has led to a massive expansion of biochemical research, and consequently biochemical knowledge. The progress achieved in protein chemistry since the turn of the century may perhaps be illustrated most strikingly in connection with the determination of the amino-acid composition of proteins: at the time of *E. Fischer* a quantity of 100 g was necessary for study of the amino-acid composition and the analysis required one month; nowadays, from 10-20 μg of a protein *hydrolysate* a high-pressure amino-acid analyser will give the result within 70 minutes, with a limiting error of $\pm$ 1%.

These new methods have enabled the rapid isolation of the most varied proteins and the establishment of their structures. Choice of the best method to use depends on the properties of the proteins, and on the aims of the work. In planning protein-analysis work it must always be remembered that different methods may give very different answers.

It would be impossible to systematise the methods for protein isolation and analysis on the basis of classification of the proteins, because there is no such classification of general validity. Whatever the basis used-the constituents, the biochemical and physiological roles in the organism, the solubility, etc.—there is always some degree of overlap between groups. It follows that there can also be no generalised guide to the methods for preparation and investigation of the proteins within the different groups. The methods used to isolate given proteins may also differ according to the origin of the proteins. For instance,

Table 1.1

Material for examination	Characteristic data, features	Preparative processing: Aim	Preparative processing: Manner	Analytical examinations applicable
Native plants and animal organs	Plant and animal tissues with high water content	Extraction and possibly isolation of component proteins	Application of homogenization, dissolution, concentration and separation methods	Nitrogen or protein determination, amino-acid determination, examination of homogeneity with electrophoresis and chromatographic methods, molecular weight determination
Physiological solutions, tissue fluids	Solutions with low protein and free amino-acid concentrations	Detection, extraction and possibly isolation of component proteins, determination of free amino-acids	Application of concentration and separation methods	Nitrogen or protein determination, amino-acid determination, free amino-acid determination, examination of homogeneity with electrophoresis and chromatographic methods, molecular weight determination
Fermentation industry products	Fermentation fluids containing dissolved and insoluble proteins, suspensions, fermented feeds	Obtaining each component protein, examination of utilization of partially decomposed proteins, detection and determination of free amino-acids and amino-acid destruction	Dependent on the technological objectives	Nitrogen or protein determination, amino-acid determination, free amino-acid determination, examination of components with electrophoresis and chromatographic methods, molecular weight determination

Contd. Table 1.1

Contd. Table 1.1

Material for examination	Characteristic data, features	Preparative processing		Analytical examinations applicable
		Aim	Manner	
Food and feed proteins	Agricultural and industrial products	Determination of protein content and amino-acid composition, examination of solubility and tendency to undergo lysis	Generally standard prescriptions for the individual preparations, depending on the practical use	Nitrogen and amino-acid determination, possibly solubility study, examination of composition of component proteins, and of molecular weight proportions
(a) Isolated proteins, hydrolysis and reaction products (b) Synthetic peptides	Biochemical products Products of preparative chemical synthesis	Chemical and biological identification, establishment of homogeneity, structural examination in the case of research products	Examination of solubility and of stability and physical properties of solution, with electrophoresis and chromatography, examination of homogeneity	Determination of physico-chemical characteristics, determination of nitrogen and amino-acid contents, partial hydrolyses, determination of terminal amino-acids and amino-acid sequence

because of the different consistencies of the starting materials, in the cases of soybean, pancreas and egg different methods must already be employed in the first step (the extraction) in the isolation of a particular protein such as trypsin inhibitor.

In protein analysis, the object is not usually establishment of the chemical structure. In the study of plant breeding, for example, if the object is the knowledge of only one of the components of the seeds (e.g. total nitrogen content, or a historic, or the nutritionally important methionine), a complete protein analysis is not necessary. The choice of method is therefore influenced jointly by the properties of the raw material and of the proteins to be investigated, and by the purpose of the analysis. Some insight into protein analysis is provided by a scheme which attempts to systematise the most common starting materials on the basis of their origin.

Since proteins are products of biosyntheses proceeding in plant or animal organisms, the native plant and animal tissues, physiological solutions and tissue fluids form the starting materials for protein examination, which may range in complexity from the simplest determination, that of nitrogen content, to the most complex problem, elucidation of the structures of the proteins and peptides isolated from them. The experimental work will correspondingly be extremely varied. The nitrogen content may be deter-mined with only a few operations, but the result may not reveal very much. In contrast, isolation of a biologically active protein necessitates a whole series of difficult separations.

The last type of materials listed in Table elsewhere in this chapter consists essentially of the product of these purification and separation operations: an isolated protein or peptide, as proved by purity tests, which is suitable for the determination of its physical, chemical and biological characteristics.

Fermentation industry products, together with food and feed proteins (which deserve special mention for economic reasons), do not usually require a full analytical examination in everyday practice. To establish the enzyme-output of an industrial-scale fermentation and the biological activity of the product, it is generally unnecessary to perform the entire series of preparative separations. Similarly, the nutritional value of food and feedstuffs can be estimated from their nitrogen content, amino-acid composition and readiness to undergo lysis, without extraction of the component proteins.

In research into protein structures, the isolation steps may sometimes be omitted, as shown by the elucidation of the insulin sequence, but then great ingenuity is needed, such as that shown by Sanger in his Nobel Prize work. The isolation of proteins can be divided into two steps: *preparation* and *separation.*

The *preparation* steps (including mechanical disintegration, homogenisation, suspension, dissolution, sedimentation, etc.) are determined not only by the analytical objectives, but also by the mechanical, physical and chemical properties of the starting material.

Before the protein isolation, the material homogenised or extracted from the starting material is purified by ultracentrifugation, fractional precipitation, filtration, etc., depending on its nature.

The *separation* and isolation of the individual component proteins (peptides) are done on the more or less purified solutions of the proteins, at the appropriate concentration. The separation may be based on differences in some fundamental physical property of the proteins: the sedimentation rate, mass, size, electric charge, hydrophilic nature or isoelectric point of the protein molecules.

With the development of isoelectric focusing (IEF), the separation techniques have undergone considerable modification. Until very recently, gel-filtration was used to achieve separation on the basis of molecular mass. Isoelectric focusing gives direct knowledge of the isoelectric points of the component proteins and hence indicates the optimum conditions for further separation or isolation, on the basis of the character of individual proteins.

Figure elsewhere in this chpater outlines a flow-chart for protein purification; it illustrates various possibilities for combining individual techniques.

The following comments may be made on the individual steps:

1. With analytical PAG IEF in the pH interval 3-10, the pI values of the component proteins are obtained, which can serve for identification or confirmation of identity.
2. On the basis of the IEF results from 1, the protein(s) of importance can be obtained by preparative flat-bed IEF with an ampholyte in a narrow pH range corresponding to the pI value(s).
3. By the simplest PAGE, electrophoresis in homogeneous 7.5% PAG, information is obtained on (a) the purity of the fraction; (b) its approximate molecular mass; (c) in the case of several components, the approximate amounts of the individual fractions.

4. Further information is obtained on the molecular mass(es) of the protein(s).
5. From comparison of the SDS-PAGE results and those of the previous two PAGE operations 2 and 3, it may be decided whether the protein is present as a monomer or a polymer.
6. Accompanying proteins (with similar pi values, but different molecular masses) in the band acquired from 2 are separated from the desired material by preparative gel-filtration with the appropriate Sephadex, on the basis of 3, 4 and 5. The success of this step depends on correct selection of the Sephadex (G-75, 100, 150).
7. Repeated purification of the detected fractions on the basis of their p*I* values, in the narrow pH range used in 2, with a restricted electrode distance, to separate the desired protein from those protein fractions not removed by the gel-filtration.
8. Analytical IEF in PAG with an ampholyte giving a narrow pH range, to confirm the results of 7. Initially the longest possible gel-field is used, then a restricted electrode distance. If isolation is successful, the desired material is eluted from the flat-bed used in 7 and the ampholyte removed by gel-filtration.
10. 12, and 13. Preparative ion-exchange chromatography on the basis of the charge of the protein, which depends on the pH of the solvent; a cation-exchanger is used for positively-charged proteins dissolved at a pH below the isoelectric point, and an anion-exchanger for negatively-charged proteins dissolved at a pH above their isoelectric points.
11. 14, and 15. Preparative chromatography of uncharged molecules by partition or adsorption chromatography.
16. As for 8.
17. With affinity chromatography it is possible to separate the substances according to their biological function.

The analytical methods used are employed both for identification and determination, and should preferably be applied after each major separation step to assess the efficiency. Examination of the starting material yields useful information on the choice of the amounts and proportions to be used. The efficiency of dissolution can be checked by determination of the amount of protein dissolved, e.g. by nitrogen

Fig. 1.1

1. Analytical PAG IEF, pH 3-10

2. Preparative flat bed IEF, within the optimum interval for the fractions to be separated

Pre-examinations for further purification:

3. Homogeneous analytical PAGE

4. Gradient analytical PAGE

5. SDS analytical PAGE

Further purification on the basis of the net charge of the proteins

7. Preparative flat-bed IEF on the necessary fractions

8. Restricted analytical IEF in PAG, in the optimum pH interval

9. Special isolation methods (e.g. affinity chromatography)

Further purification on the basis of molecular mass and charge

6. Preparative column gel-filtration

10. Ion-exchange preparative chromatography of charged proteins

11. Uncharged proteins

12. Cation-exchange chromatography of positively-charged proteins

13 Anion-exchange chromatography of negatively-charged proteins

14. Adsorption chromatography

15. Partition chromatography

16. Restricted analytical IEF in PAG, in the optimum pH interval, or

17. Affinity chromatography

Abbreviations: I EF = isoelectric focusing
PAG = polyacrylamide gel
PAGE =polyacrylamide gel electrophoresis SDS=sodium dodecylsulphate

determination, the Lowry reaction, the biuret reaction, or some specific reaction of the solute, such as an enzymic assay.

From the protein and biological activity determined in the course of extraction of biologically active substances, the purity achieved can be assessed. It is useful to measure the following quantities and tabulate them:

Protein concentration, mg/ml

Volume, ml

Total protein, mg

Total activity units

Specific activity, units/mg

Yield, %

Purity, %

One of the most important task's is to establish whether a fraction exhibiting a single chromatographic peak or electrophoretic band corre-sponds to a single component or an unresolved mixture. This can be done by using essentially the same techniques as for the preparative separation: sedimentation, electrophoresis, gel filtration and chromatography.

Different authors have different ideas about characterisation and structural examination of isolated proteins. We will illustrate this from the proposals of two authors.

Jolles classifies structural methods into two main groups. The first contains the classical structure research methods in ten steps.

1. Determination of the amino-acid composition.
2. Determination of the molecular weight.
3. Qualitative and quantitative determination of the N-terminal amino-acid and sequence.
4. Qualitative and quantitative determination of the C-terminal amino-acid and sequence.
5. Separation of the peptide chains, if the molecule consists of several.
6. Specific cleavage of every peptide chain into smaller peptide fragments, preferably with trypsin.
7. Purification and separation of the fragments.
8. Determination of the structures of the fragments. 9. Compilation of the fragments.
10. Establishment of the positions of the disulphide bridges.

The second group contains the three methods for direct establishment of the structure.

1. X-ray structural analysis, with which the stereostructure of the protein can be derived from knowing the primary structure.
2. Totally automated Edman decomposition.
3. Mass spectrometry, for identification of peptides.

Rothfus gives the following characterisation of proteins: titration, hydrogen exchange, enzymic cleavage, reactions with bifunctional reagents, dye-binding, sedimentation, light scattering, viscosity measurement, osmometry, dialysis and diffusion, ultraviolet spectroscopy, infrared spectroscopy, optical rotatory dispersion, circular dichroism, fluorescence measurements, electron paramagnetic resonance spectrometry, nuclear magnetic resonance spectrometry, differential thermal analysis, X-ray crystallography, sequence analysis.

The listings by both authors are subjective, and neither can lay claim to exclusive validity. In any event they provide a picture of the complexity of characterising proteins and show that at this level a significant proportion of protein analysis no longer belongs only to the sphere of biochemistry.

In this book we are confined to presentation of only the possibilities of application of protein analysis, and a fundamental technical description of the individual methods. This is supplemented in the appropriate chapters by a list of those monographs and technical books which may satisfy the needs of those interested in the theoretical basis of the methods, and in the more complex aspects of protein research.

2

Protein Aggregation

Protein denaturation and aggregation are frequently occurring phenomena, long recognised as common processes. The aggregation under consideration is very different from precipitation of a native protein at the isoelectric point or upon salting out, which can be reversed by dissolution under appropriate conditions. The aggregates are formed from the nonnative protein upon *intermolecular* interactions, which compete with intramolecular interactions. These aggregated species are not in equilibrium with soluble species, complicating the experimentation. Aggregation has been reported to occur during the in vitro refolding for monomeric as well as oligomeric proteins, lowering the refolding efficiency.

Aggregates can also accumulate during the in vivo folding of nascent polypeptide chains synthesised in prokaryotic and eukaryotic cells. The overexpression of genes introduced in foreign hosts often results in aggregated nonnative proteins termed "inclusion bodies," introducing a serious limitation in the production of recombinant proteins. Such a problem remains critical in the development of biotechnology. In any case, protein aggregation is an offpathway event associated with protein folding or protein association, which results from kinetic competition between proper folding and a side reaction. Such competition, which arises from an early folding intermediate producing misfolded species, can lead to the formation of aggregates.

PROTEIN AGGREGATION IN IN VITRO REFOLDING

The refolding of an unfolded protein has been recognised to be a spontaneous process under in vitro conditions. Although it was shown by Anfinsen in the 1960s that the threedimensional structure of a protein is determined by its amino acid sequence, the mechanisms involved in the folding process are not well understood. Solving the folding problem, this "second translation" of the genetic code, is of great importance, not only for understanding the mechanisms by which a polypeptide chain acquires its threedimensional and functional structure, but also for progress in biotechnology.

Monomeric Proteins

The formation of aggregates has been reported for proteins with disulfide bonds as well as for proteins devoid of disulfide bonds. Thermal unfolding of proteins is frequently accompanied by the formation of aggregates and therefore behaves as an irreversible process. Thermal unfolding occurs at temperatures that vary widely according to the protein, since the temperature of optimum stability depends on the balance between hydrogen bonds and hydrophobic interactions. Generally, the products of thermal denaturation are not completely unfolded and retain some structured regions. The unfolding of small proteins induced by urea or guanidine hydrochloride is generally reversible under appropriate conditions (i.e., pH, ionic strength, temperature, and low protein concentration).

The presence of covalent crosslinks, such as disulfide bridges, often requires the addition of small amounts of a reducing reagent or a redox mixture to prevent or to correct a wrong pairing of the halfcystines, and further aggregation. The "reshuffling" of disulfide bridges during the refolding of a protein takes place through a series of redox equilibria according to either an intramolecular exchange or an intermolecular exchange. It is possible to prevent misfolding and aggregation for proteins with disulfide bonds by monitoring the refolding in rigorously determined redox regeneration mixtures.

The formation of aggregates has also been observed during the refolding of monomeric proteins devoid of disulfide bridges whose polypeptide chain is folded in several structural domains. At a critical concentration of denaturant, which often is very close to that of the end of the unfolding transition, the denaturant induces an apparent irreversibility in the unfoldingrefolding transitions. This phenomenon is strongly dependent on the protein concentration and on the time of exposure to the denaturant, reflecting an aggregation process. This apparent irreversibility is also temperature dependent. It has been attributed to side reactions occurring through intermediates on the folding path from which incorrectly folded species are formed. These latter species sometimes, but not always, associate into aggregates through hydrophobic interactions. The nature of the intermediate that generates the incorrectly folded species and/or the aggregates remains a matter of discussion. The outcome of many investigations on protein folding indicates a great variety of early transient intermediates, formed in a fraction of a second. However, the emergence of the final tertiary and functional structure is a late event and is the ratelimiting step in the process. One of these early intermediates, named the *molten globule,* was characterised as having a native like secondary structure with a high flexibility and a tertiary structure not yet formed. In this molten globule, hydrophobic surfaces remain exposed to the solvent, favoring the formation of aggregates. Although this intermediate has been suggested to be a good candidate for the formation of aggregates, there is no conclusive proof that the aggregates arise from the molten globule state. To overcome this complication in the refolding of a protein, one must have very precise conditions of pH, protein concentration, temperature, ionic strength, and redox mixture, when necessary. These conditions differ from one protein to another.

Oligomeric Proteins

With oligomeric proteins, a new degree of complexity appears in the hierarchy of protein structure and folding. It is generally accepted that the early steps in the acquisition of functional structure by oligomeric proteins are practically identical to those occurring in the folding of monomeric proteins. Within a subunit, it is reasonable to assume that first some stretches of structure are formed in several parts of the molecule; these interact to generate subdomains, which then associate to produce structural domains, and the folded monomer is formed upon domain pairing with a nativelike structure. In a last step, subunit association and subsequent conformational readjustments yield the native and functional oligomeric protein. For the process to be completed, the subunit interfaces must be correctly recognised. The refolding of several oligomeric proteins, dimers and tetramers, has been extensively studied by R. Jaenicke and his group, who have described the overall process of formation of an oligomeric protein as a sequence of firstorder (folding) and secondorder (association) reactions. For oligomeric as well as for monomeric proteins, side reactions could occur in an offpathway, leading to the formation of aggregates from misfolded species. Once again, there is kinetic competition between folding/association and aggregation, depending on the experimental conditions.

Methods to detect and Characterise Protein Aggregates

The detection and characterisation of aggregates represent important aspects in folding studies. The most usual methods to determine the association state of a protein, available in any laboratory of biochemistry, are gel permeation and sodium dodecyl sulfate polyacrylamide gel electrophoresis with and without crosslinking. The detection of aggregates also can be monitored by other hydrodynamic methods such as analytical ultracentrifugation and viscosity measurements. Classical light scattering is a convenient method to detect and characterise aggregates and to give information on their size. Quasielastic light scattering is a dynamic method that can be used to determine macromolecule diffusion coefficients as a function of time (i.e., to follow the kinetics of aggregation). Electron microsc-opy is often used in the characterisation of aggregates in the in vivo folding. In particular, the shape and dimensions of inclusion bodies have been determined by this method.

Rules of Protein Folding Deduced From in Vitro Refolding Studies

Although the mechanism of protein folding is not completely understood, the great number of in vitro folding studies has provided a significant corpus of information allowing the deduction of some main rules.

1. The folding of a polypeptide chain is determined by the amino acid sequence in a given environment. Under appropriate conditions, it is a spontaneous process.

2. The structure of a native protein is thermodynamically controlled; its stability is not very high, as estimated by the variation in free energy AG from the unfolded to the folded state, whose value varies between 5 and 15 kcal/mol according to the protein.

3. The folding pathway involves partially folded intermediates, which are more or less populated according to the protein and the conditions. Their formation is kinetically controlled. Some of these prefolded intermediates can evolve into misfol-ded species, which may or may not result in aggregation depending on the conditions. There is kinetic competition between proper folding and misfolding and subsequent aggregation.

PROTEIN AGGREGATION DURING\ IN VIVO FOLDING

Folding in the Cellular Environment

The main concepts concerning protein folding were developed

from in vitro studies, and it is accepted that the in vitro refolding process is a good model for understanding the mechanisms by which a polypeptide chain reaches its native state in the cellular environment. However, the recent discovery of molecular chaperones has led to a reevaluation of this hypothesis. Indeed, the environmental conditions within cells are markedly different from those used in the in vitro refolding studies. Different factors can affect the in vivo folding pathway, the synthesis, the co and posttranslational covalent modifications, and the localisation of the mature protein. Protein aggregation can occur during the in vivo folding of a nascent polypeptide chain. The phenomenon is often associated with pathological states and can occur in cells either carrying certain genetic lesions or grown under stressful conditions. The overexpression of active recombinant proteins in foreign hosts frequently results in the accumulation of insoluble aggregates, called "inclusion bodies." In biotechnology, the solubilisation and proper folding of these aggregates often represents a crucial step in the production of the active product.

The Formation of Inclusion Bodies

Occurrence of inclusion bodies

In cells, inclusion bodies appear as amorphous aggregates clearly separated from the rest of the cytoplasm; they form a highly refractile area when observed microscopically. A great variety of experimental approaches indicates that the formation of inclusion bodies results from partially folded intermediates in the intracellular folding pathway of proteins, not from the totally unfolded or native proteins. This is particularly well illustrated by the folding pathway of tailspike endorhamnosidase from *Salmonella* phage 22, studied by the group of J. King, one of the few systems for which the in vivo pathway has been compared with the in vitro refolding pathway. In the production of engineered proteins, the formation of inclusion bodies generally appears to be a disadvantage, since it requires the aggregates to be dissolved in denaturant and subsequent refolding of the protein. But when the active product can be recovered in a sufficient yield, certain advantages may accrue. Indeed, aggregation generally prevents proteolytic attack, with the exception of pyruvate kinase and bovine pancreatic trypsin inhibitor, expressed in *Escherichia, coli* which coaggregate with a protease in inclusion bodies. The formation of inclusion bodies is also an advantage for the production of proteins that are toxic for the host cells. Furthermore, inclusion bodies contain a great quantity of the expressed protein.

Characteristics of inclusion bodies

The characteristics of the aggregates depend on how the protein is expressed. Different sizes and morphologies have been observed. Generally, inclusion bodies appear as dense isomorphous aggregates of nonnative protein separated from the rest of the cytoplasm, but not surrounded by a membrane. They look like refractile inclusions and when large enough can be easily recognised by phase contrast microscopy.

Machanisms of inclusion body formation

The conformation of the polypeptide chains in inclusion bodies is not known. The particulate character of the aggregates and their properties of light scattering prevent the use of spectroscopic methods. For proteins containing disulfide bridges such as prochymosin, the formation of incorrect intermolecular disulfide bridges can partially account for the formation of aggregates. But inclusion bodies have been also observed for proteins devoid of disulfide bridges. In general, the aggregates result from noncovalent intermolecular interactions, mostly hydrophobic interactions, and they can be dissolved using high concentrations of denaturant. Nevertheless, comparison of in vitro and in vivo studies of the aggregation phenomenon has led to the conclusion that aggregates do not result from the association of completely unfolded molecules, but are generated from early intermediates of the folding pathway.

Strategies for refoldinginclusion body proteins

The main problem in the production of recombinant proteins is the low refolding yield of the biologically active product. The recovery of the active protein involves a succession of several steps, including cell disruption and separation of the insoluble material, solubilisation of inclusion bodies in the presence of strong denaturants, refolding, and purification. To maximise refolding yields, stabilizing reagents such as sugars, alcohols, and polyols (including sucrose, glycerol, polyethylene glycol, and isopropanol) may be added. Conversely, the use of solubilizing reagents may be helpful in dissolving the aggregates. Another important factor in the refolding process is the rate of removal of the denaturant. Since there is kinetic competition between correct refolding and the formation of aggregates from an intermediate, conditions that favor folding over the accumulation of aggregates must be found.

Role Of MoleCular Chaperones in Protein Folding

The recent discovery of molecular chaperones, a ubiquitous class of proteins mediating the correct folding of polypeptide chains in the cellular environment, led to a reconsideration of the mechanisms of in vivo protein folding. Ellis extended the concept of the molecular chaperone to define a class of proteins whose function is to ensure the correct folding and assembly of newly synthesised proteins through a transient association with the nascent polypeptide chain. Studies of heat shock proteins have widely contributed to the development of this concept. These proteins, induced by high temperatures, associate with proteins that are partially unfolded as a consequence of the thermal stress and thus are prevented from aggregating. Heat shock proteins reportedly have participated in the assembly of oligomeric proteins under normal conditions. These proteins have been classified according to molecular weight into three families, hsp 60 or GroEL, hsp 70, and hsp 90.

The hsp 60 chaperonins

The chaperonin family is a class of related proteins found in all bacteria and mitochondria and also in chloroplasts. There are two types of chaperonins, cpn 60 and cpn 10 (in *E. coli,* GroEL and GroES, respectively), with subunit molecular weight of 60 and 10 kDa, respectively. These chaperonins are constitutive proteins whose amount increases under stress conditions, such as heat shock, bacterial infection, or an increase in the cellular amount of unfolded proteins. Although chaperonins efficiently facilitate the correct folding and assembly of the polypeptide chain, preventing the formation of aggregates, they are unable to reverse the aggregation process. The dissociation of the complex is dependent on the hydrolysis of ATP for several proteins. ATP hydrolysis is not required for the release of certain proteins however, since a nonhydrolyzable analogue is capable of promoting the release of the protein. These results clearly indicate that the binding, folding, and release of the folded proteins are distinct steps.

The hsp 70 chaperonins

The heat shock protein family hsp 70 consists of several structurally related proteins, all binding ATP with high affinity. Representatives have been found in bacteria, mitochondria, and chloroplasts, and also in eukaryotic cells. Several are strictly inducible by heat shock or other form of cell stress. Others, referred to as cognate heat shock proteins, are constitutive. It is likely that the constitutive heat shock proteins play a role in posttranslational transmembrane targeting of proteins to cellular organelles, and also in the disassembly of clathrin cages. In yeast cells, the cytosolic hsp 70 proteins are required for membrane translocation, and it was suggested that they are responsible for ATPdependent unfolding of the preprotein before translocation.

The hsp 90 chaperonins

Members of the hsp 90 class are also considered as molecular chaperones; indeed the chaperonin function of hsp 90 has been proven by in vitro experiments and has a role in steroid receptors biosynthesis.

The rules of protein folding and the concept of molecular chaperones

By their transient association with nascent, stressdestabilised, or translocated proteins, molecular chaperonins have a role in preventing improper folding and subsequent aggregation. However, they are unable to interact with native proteins, or to refold aggregates once formed. Furthermore, they do not carry information capable of directing a protein to assume a structure different from the one dictated by the polypeptide amino acid sequence. Therefore, they assist the in vivo folding without violating the main rules of protein folding as deduced from in vitro experiments. The molecular details of chaperoneprotein interactions are not yet known. The development of in vitro folding and assembly physicochemical studies with highly purified proteins and molecular chaperones will allow substantial progress in the understanding of in vivo protein folding.

Other proteins that assist in vivo folding

Other such proteins that assist in vivo in the correct folding and assembly of proteins have been described. The two enzymes, protein disulfide isomerase and *peptidyl-prolyl-cistransisomerase*, cannot be regarded as molecular chaperones.

3

EXAMINATION TECHNIQUES

Depending on the heterogeneity of the material, the mass of the sample may range from a few grams to several kilograms. In any event it is many times the quantity taken for the actual analytical examination. As an example, agricultural feed (e.g. roughage, hay) may be the most *heterogeneous* material to be examined in laboratory practice. It must be made as representative as possible, and many 200-250 g samples must be taken (at least 20 samples per 25 tons of unpressed hay, or per 50 tons of pressed hay) and combined to give average samples of about 5 kg. In the course of the laboratory preparation, therefore, there is of necessity a second sampling (we may call this internal sampling), but this must be preceded by rough homogenisation of the primary sample, without which the original sampling would lose its significance. The correct procedure then is to bring the whole of the primary sample into a state suitable for homogenisation, and to employ laboratory equipment or semi-plant-scale machines to homogenise it in the manner most appropriate for the purpose. For example, solids that can be ground should be reduced to a grain size of at most 1-2 mm.

For umpire, *referee* or *official* examination, the entire analysis must be carried out in the prescribed agreed manner. Since these procedures are readily available in official handbooks, it is not necessary to describe them here. We are primarily concerned with the analyses connected with various fields of research and application in agriculture and industry, the methods for which are selected by the specialist, according to the objectives, number of analyses, accuracy required and the technical level of the laboratory.

In practice there are three possibilities for decreasing the scatter in the results of an examination:

(a) increasing the quantity of material taken for analysis;

(b) increasing the number of replicate determinations;

(c) improvement of the homogeneity of the average sample.

Increasing the amount of material taken precludes use of a whole series of modern methods that have been developed for micro and semimicro-scale work, and at the same time increases the costs and possibly the time of the analysis. This approach is therefore used only in the opening-out phase (dissolution, destruction, ashing) and the analysis is subsequently continued on an aliquot. The optimum number of replicates demanded by statistics is usually much larger than the number dictated by practical considerations such as time, equipment, labour-force, costs, etc. Frequently, therefore, the best solution is to improve the homogeneity of the sample. Thus, in the internal sampling, a solid sample must be ground to such a fineness that a quantity of 0.1-1 g will be adequately representative of the whole. According to practical experience, this requires a particle size range of 200-500 Pin. The comminution must be done so that the temperature does not exceed 45°C during the homogenisation. It is advisable to homogenise moist materials in two steps, if the starting material is *heterogeneous* and in the kg range. It is very important to mix the samples again just before portions are taken for analysis, to avoid introduction of *heterogeneity* by sedimentation.

Table 3.1 : Standard deviations of nitrogen analyses as a function of the nitrogen content of the sample.

Nitrogen, %	*R*, %	*S*,%	*v*, %
1	0.036	0.032	3.2
3	0.057	0.051	1.7
5	0.062	0.055	1.1
15	0.086	0.076	0.5

The smallest quantity of sample that should be taken is governed not only by the *homogeneity* of the material, but also by the sensitivity of the analytical method. Further, the scatter in the results depends not only on the systematic error in the analytical procedure but also on the concentration of the determinand in the sample to be analysed. For example, the estimated standard deviation for duplicate Kjeldahl nitrogen determinations was found to increase with nitrogen content, but the relative standard deviation (coefficient of variation) decreased.

R is the mean range (difference between duplicates) for a large number of samples, S is the estimated standard deviation calculated from the relation $S=Rk$, where n is the number of replicates, and ν is the coefficient of variance given by

$$\nu = \frac{S}{x}.100$$

It can be seen that the standard deviation increases with increasing nitrogen content, the coefficient of variation decreases.

Table 3.2a : Values of the coefficients k_n for estimation of the standard deviation, according to Dean and Dixon.

n	k_n
2	0.886
3	0.591
4	0.486
5	0.430
6	0.395
7	0.370
8	0.351
9	0.337
10	0.325

If the procedure is particularly difficult or time-consuming there will seldom be more than duplicate analyses performed. If the results agree closely enough the average may be used; otherwise a third analysis must be performed to decide which of the results is more nearly correct. However, the question arises as to what constitutes a close enough agreement. With only two results we cannot proceed according to the strict methods of mathematical statistics; instead, an estimation procedure must be employed. The maximum difference R_{max} between the results of duplicate analyses may be calculated from

$$R_{max} = bR$$

where the *b* values are taken from Table, for probability levels P = 10, 5 and 1%, and *R* has the same meaning as before.

Thus for the Kjeldahl nitrogen determination, if the nitrogen content of the samples is about 3%, the value of R is 0.057%, and for a probability level P= 5% and n = 2 repetitions, we have b = 2.46.

Hence

$R_{max} = 2.46 \times 0.057\% = 0.14\%$

Suppose that the two results are 3.19 and 2.96%. The difference is 0.23%, which is larger than the permitted R_{max} =0.14%. One of the results therefore involves a gross error; and accordingly a third

Table 3.2b : Values of b for calculation of the maximum range permitted for probability levels P = 100, 5% and 1%, in the case of n replicate analyses.

n	*h*		
	P=10%	P=5%	P=1%
2	2.06	2.46	3.23
3	1.71	1.96	2.43
4	1.57	1.76	2.14
5	1.50	1.66	1.98

analysis must be done. Suppose it gives 2.92%. In this case, the difference 2.96-2.92=0.04% is smaller than the calculated 0.14%, and the two values can therefore be averaged. The value 3.19%, on the other hand, must be rejected, as involving a gross error.

Homogenisation of Samples

For sample preparation, various types of apparatus may be used in one or more steps to attain the required degree of comminution and *homogeneity*. This equipment uses percussion, shearing, cutting and friction, which produce various amounts of heat, depending on the resistance of the material, the construction of the apparatus and the rate of the operation. The requirements are similar to those to be met in drying. The temperature must not be so high that the examination material undergoes damage. This can be achieved by appropriate choice of apparatus, correct rate of addition of the sample, and by cooling if necessary.

A further requirement is that the apparatus should be chemically resistant to the materials in the sample and another very important factor is that it should be easy to clean.

Dry solid materials

Comminution of a large mass of solid samples such as dry plant materials is best begun with slicing machines and continued with a universal mill. The rotor of a universal mill operates at 3000-15000 rpm and is of various types such as a rotating cross, a swing hammer, a series of hammers or a serrated disc. The stationary part of the mill may be smooth or serrated, and there may be a sieve over a greater or smaller portion of the cylinder jacket. The use of disc grinders cannot be recommended, as these are difficult to clean and it is difficult to avoid the heat-effect associated with the shearing.

Ball and roller mills may be used for the comminution of brittle, relatively *inelastic* materials containing fibres. In the case of naturally-occurring materials the possibilities of their use are limited. The comminution is slow, the heat-effect is high (although this can be counteracted somewhat by external cooling), it is not easy to take out the sample, and the apparatus is difficult to clean. However, there can be no doubt that a very high degree of *homogeneity* can be achieved with such equipment, with reduction to an average particle size of 0.5 mm.

The efficiency of comminution of dry solid materials may be checked by classification with a set of sieves shaken *mechanically*. It must be emphasized that the sieving is intended primarily to be a means of controlling the comminution process. If it is used for grading, it must always be taken into account that the sample (which should be uniform for analysis purposes) will have been resolved into fractions on the basis of particle size, and that these fractions will differ in *composition* because each component will have its own particle size distribution, and these will all be different. In certain cases, however, e.g. with milling-industry products, the aim may well be to examine the fractions with different *particle* sizes.

Moist solid materials

The objective is the production of homogeneous suspensions with various consistencies. Depending on the original moisture content of the material and on the aim and means of the further processing, the suspensions may be prepared with or without the addition of a suspending agent. The suspensions are prepared with *homogenisers*. The apparatus known as a '*Waring blender*' comes in two different types. In the first, the sample is placed in a ribbed glass or metal vessel, various amounts of water or some solution are added, and comminution is effected by two-armed, cross- or star-shaped knives, generally rotating in several planes on a common spindle, at 10 000-25000 rpm. The ribbing helps to prevent the mixture from flowing together with the knives, and the material is comminuted by the knives as a consequence of the resulting eddy resistance. The homogeniser vessel is usually surrounded by a cooling jacket, to prevent overheating. The cutting knives of the other type rotate in a serrated *castellated sheath*, the apparatus operating on the principle of centrifugal turbines.

Larger samples may be homogenised in a vessel into which a coarse suspension is fed so that it flows through at a rate corresponding to the efficiency of *homogenisation*.

Homogenisers of '*Potter-Elvehjem*' type are suitable for the preparation of cell-free material. The essence of these homogenisers is that a ground-glass or *Teflon piston* is moved, either manually or mechanically, in a concentric ground-glass tube. The material passing between the two surfaces attains a very fine degree of dispersion as a consequence of the friction, and is homogenised so as to be cell-free. This type of apparatus is suitable only for the homogenisation of small samples which have previously been coarsely dispersed in some other *homogeniser*.

It must be noted that very good homogenisation can also be obtained by grinding the material with a mortar and pestle, by the traditional simple method, with the addition of powdered sand or glass.

For preparation of the homogeneous suspension, water or some aqueous solution is used as suspending agent. The suspension medium may be a buffer solution, or salt or sugar solutions of various concentrations. If the aim of the homogenisation is to obtain intact cells in the suspension, then solutions of isotonic concentration are necessary. *Detergents*, *chelating agents*, *anti-oxidants*, *enzyme-inhibitors*, etc. may be employed in the homogenizing solution.

Ultrasonic disintegrators may be used to advantage, mainly for extraction of the proteins of micro-organisms. Instruments manufactured for laboratory use operate in the frequency range around 20 kHz. With these, and a vessel and vibrator of good geometry, an energy concentration of as much as 50 W/cm^2 can be achieved. In the ultrasonic instruments, the output of a high-frequency generator is transformed by a *piezoelectric* cell into mechanical vibration, which is transmitted to the liquid by a *vibrator* made from a metal (e.g. titanium) resistant to mechanical and chemical effects. The vibration causes high pressure and suction forces in the liquid and the formation of extremely small bubbles in the interior of which the pressure is very low. This *phenomenon* is known as *cavitation*. Since the vibration *frequency* is very high, the bubbles have a short life-time, less than 10 μsec. As a consequence of the disappearance of the *bubbles*, very high *ultrasonic* pressures develop in the liquid. The liquid comes into motion, the velocity of the particles attaining 30 m/s, and their acceleration 10^5 g. This *mechanical* effect is capable of rupturing the cell wall, and thus the internal content of the cell becomes free. Since the different *biomembranes* differ in resilience, ultrasonic disintegrators may be used to carry out fractionated cell rupture according to a programme developed by previous investigation with the particular instrument and materials. Though the method is most suitable for obtaining cell proteins, various effects must be reckoned with. Ultrasonic vibration is accompanied by a thermal effect, and without satisfactory cooling the material may be overheated. There is also a *depolymerising* effect, and thus aggregated protein molecules are disaggregated. It may readily be observed that the electropherograms of protein solutions display changes after ultrasonic treatment of the solutions: certain fractions disappear, and new ones arise. In liquids, the ultrasonic waves are decelerated over a short distance, and uniform treatment can be attained only in a small volume. A flux regulator can be used to achieve uniformity of treatment.

Drying of the Material

The samples to be examined may be liquids or solids of biological origin, that are *heterogeneous* to various extents; by virtue of their composition and moisture content, they may be readily liable to change, either in themselves or as a result of *microbial* processes. The sample preparation therefore has to be considered in conjunction with storage of the samples. Until dried or processed, liquid samples or solids with a high *moisture* content must be stored in deep-freezers, in which they can be kept at a temperature of –25 °C or below. A temperature of –60 °C is necessary for the complete freezing of animal tissues.

The technical equipment for mild drying consists of vacuum-driers and *lyophilizing* apparatus.

The vacuum-drying ovens used in the laboratory can be heated in the range 40-150°C and are pressure-tight; they are *cylindrical* in shape, and fitted with a glass-windowed door, through which the state of the sample may be inspected. If the water pressure is favourable and constant a water-pump can be used to reduce the pressure, but a rotary oil-pump with gas-ballast valve is generally used. In such equipment, *bulkier* samples may be dried at 40-60°C.

Lyophilisation, the most expedient means of removing water from thermally labile biological materials, is used in several phases of the work in a protein laboratory. Lyophilisation, or freeze-drying, removes water by the sublimation of ice from the frozen sample at a temperature in the range between 0 and –60 °C, at a pressure of between 13.3 and 0.13 Pa (0.1 and 0.001 mmHg). Apart from the low temperature, the method has the great advantage (particularly in the case of colloidal solutions) that dissolution of the *lyophilised* preparations is much easier than with any other drying procedure.

Drying by vacuum-evaporation

A frequent task in biochemical laboratory practice is concentration of dilute aqueous solutions, and sometimes they must be evaporated to dryness. In almost all cases this must not be done by simply boiling off the water at *atmospheric pressure*, especially with protein, peptide and amino-acid solutions, for which a lower temperature is necessary.

Vacuum-evaporation and *lyophilisation* are very suitable for this purpose, vacuum-evaporation being used mainly for peptides and amino-acids, and lyophilisation for proteins. The essential difference between the two procedures is whether the water is evaporated from the liquid phase, or sublimed from the solid phase at a temperature below freezing-point.

Table 3.3 : Vapour pressure above water and ice.

Temperature °C	Pressure Pa (mmHg)	Temperature °C	Pressure Pa (mmHg)
+60	19 918 (149.4)	0	613 (4.6)
+50	12 333 (92.5)	–10	260 (1.95)
+40	7 372 (55.3)	–20	103 (0.77)
+30	4 240 (31.8)	–30	31.3 (0.28)
+20	2 333 (17.5)	–40	12.9 (0.097)
+10	1 227 (9.2)	–50	4.0 (0.030)
0	613 (4.6)	–60	1.1 (0.008)

It is clear from table that water vapour can be sublimed from ice at a temperature lower than 0° C only at a pressure below 600 Pa (4.5 mmHg). At temperatures higher than 0° C, water vapour can be removed from the liquid phase at a pressure less than the water vapour pressure corresponding to the given temperature.

The basis of successful vacuum technique is choice of the proper pump for the job.

In simple laboratory operations, water-pumps or rotary oil-pumps are employed for filtration and vacuum-distillation. Water-pumps made of glass or plastic have the advantage that they are also suitable for the removal of corrosive vapours (e.g. *hydrogen chloride*) without the use of absorbers or cold-traps. Their disadvantages are that the performance is comparatively poor, the water-consumption high, and the attainable pressure-reduction not very high; moreover, if there is a decrease in the water pressure they readily suck back, and a suitable valve must therefore always be employed.

Rotary oil-pumps are most frequently used for laboratory operations requiring a vacuum. In the case of distillation and evaporation, it must be borne in mind that vapours condensing

Table 3.4 : Operating data on vacuum-pumps.

Type of pump	Operating range,	End-vacuum,	Suction rate
Pa (mmHg)	Pa (mmHg)		
Water-pumps	1600 (12)	Pressure of water vapour at operating temperature. Water requirements: I I of water/0.6I of air sucked off	1001/h
Rotary oil-pumps	0.13-1.3 (10^{-3} - 10^{-2})		
one-stage		0.26-13.3 (0.002-0.1)	1 - 10^3 m^3/h
two-stage		0.00133-0.0133 (10^{-5} -10^{-4})	
Diffusion-pumps	0.000133 -1.3 (10^{-6} - 10^{-2})	0.000133 (10^{-6})	0.1 -10^3 l/s
pr' ,acuum	13.3 (0.1)		

ordissolving in the oil of the pump, or forming an emulsion with it, will lower the efficiency. To remove such vapours, it is advisable to insert a cold-trap between the evacuated vessel and the pump. Even at a pressure of 0.013 Pa (10^{-4} mmHg) water vapour can be condensed in a glass or metal cold-trap placed in a Dewar flask kept at –80°C with a mixture of '*dry ice*' + alcohol. In the case of corrosive gases and vapours, traps containing suitable absorbents must be inserted. If only a medium vacuum is required, the efficiency of a rotary pump can be maintained (in spite of the effect of water vapour on the pump) if the oil is continuously exchanged during operation. The condensed vapour is removed from pumps fitted with a gas-ballast valve. Oil-pumps will give their specified performance only if *vacuum-oil* of suitable quality is used.

In selecting the size of pump to use it is practical to apply the following rough correlation:

$$t = \frac{6.5V}{S}$$

where t is the suction time necessary to attain a pressure of 133 Pa (1 mmHg), V is the volume to be evacuated, and S is the suction rate of the pump.

The vacuum applied must reduce the pressure to below that of water vapour at the desired evaporation temperature. The use of a vacuum-pump with a performance high enough to remove the vapours formed would not be *economical*. Accordingly, the vapour formed must be condensed. The efficiency of evaporation will obviously be higher, the more perfect the *condensation* of the vapour, which can be achieved by increasing the cooling surface and the intensity of cooling. A spiral condenser is generally used, cooled by mains water, but performance can be enhanced by circulation of a refrigerated coolant, e.g. 25% w/w or 23% v/v *ethylene glycol-water* mixture at 107°C.

The efficiency of vacuum-evaporation may be raised by increasing the surface area of the liquid being evaporated. This is done in rotary film-evaporators. The principle was originated by *Craig et al.*. A not too large quantity of the liquid to be evaporated is placed in a spherical or pear-shaped flask immersed in a water-bath and rotated at 50-300 rpm.

During recent years, various types of rotary evaporator have been developed. With the normal type, about 2.5 litres of water can be evaporated per hour at a pressure of 2.67-9.33 kPa (20-70 mmHg).

Small amounts of liquids can be evaporated from a test-tube or a *beaker* in a *vacuum desiccator. Phosphorus pentoxide* may conveniently be used to bind the water vapour.

In general, whatever evaporation method is employed, if the solution to be evaporated contains a volatile solvent such as alcohol, acetone, ether or a volatile acid, it is advisable to insert an absorption vessel in order to protect the oil-pump from *corrosion* and from the *deleterious* effects caused by the solvents dissolving in the oil. The cooling should be regulated so that the water vapour is condensed quantitatively. If this cannot be achieved, then a *calcium chloride*, silica gel or *sulphuric acid* water-trap must also be inserted. During the evaporation, the boiling point of the solution to be evaporated rises as a consequence of the concentration; accordingly, it is necessary to increase the vacuum or, if doing so has no harmful effects, the *temperature*. If the solution undergoing evaporation also contains electrolytes, these will be concentrated together with the material of interest. Any damaging effects of the increased electrolyte concentration must be reckoned with. In the case of evaporation of volatile acids, there will be a change in the ionic equilibrium in the concentrated solution, and hence in the pH. These effects must be taken into consideration in planning the evaporation. It is advisable to remove the excess of *electrolytes* beforehand by *dialysis* or *gel-filtration*.

Drying by lyophilisation

Lyophilisation is used for removal of the water contents of various water-containing plant and animal tissues and organs and micro-organisms, in the same way as for the evaporation of solutions of thermally labile *biochemical* substances. It is the mildest drying procedure. *Lyophilisation* of thermally labile solutions may be necessary at almost any stage in an analysis and particularly in preparative work.

Apart from fundamental *physico-chemical* and *biochemical* considerations, the lyophilisation of substances possessing structure also demands much experience, and in practice the most suitable procedure must be worked out for each material. In this connection it is especially important that the first stage of the operation, the freezing of the material, should be carried out correctly. The freezing of an aqueous solution begins with the crystalline freezing-out of ice (this also holds for tissues). In the following stage, eutectic mixtures crystallise out. The length of this stage is governed mainly by capillary effects and surface forces. The intactness of the fine structure depends on the dimensions of the crystals formed. If the temperature falls slowly, large crystals are obtained. Rapid cooling results in a *microcrystalline* structure, which damages the structure less.

In addition to resulting in large ice crystals, which mechanically damage the fine structure, slow freezing has the disadvantage that the concentration of solutes in the liquid phase increases because of conversion of part of the water into ice. This solution will be *hypertonic* relative to the cells and will cause *plasmolysis*. There will also be a large depression of the freezing point, particularly if the tissue to be frozen contains Ca^{2+} and Cl^- ions in large amounts, since the eutectic temperature for $CaCl_2$ is very low. Generally, tissues may best be maintained in the native state if a temperature of -60°C is attained as quickly as possible. In the case of protein-containing materials, it must also be considered that the *intramolecularly*-bound water cannot be frozen out, even at the lowest temperature. The retention of viability of micro-organisms during lyophilisation can be enhanced by use of protecting solutions. These usually contain two components: a *colloid* (this may be 5-20% protein) and 5-10% sugar. The purpose of the sugar is to prevent too *rapid* and *strong* a *dehydration*. With materials sensitive to oxidation it is practical to add to the protecting solution a low concentration of *ascorbic acid* or other *anti-oxidant*. The use of protecting solutions is again largely empirical because it is impossible to find a universally applicable system.

Numerous factors must also be taken into account in the freezing of *aqueous* solutions. Experience shows that at temperatures between 0 and –40°C the comparatively labile secondary, tertiary and quaternary structures of a protein, which determine the spatial conformation, do not undergo irreversible change. The stability is altered, however, by a high salt content and by a pH value different from the optimum. The presence of organic solvents such as acetone and alcohol in the solution of the protein to be lyophilised is not desirable, since it causes strong depression of the freezing point. Hence, the danger increases that the frozen solution will melt during lyophilisation and the protein will coagulate. Solvents with low boiling-points have high partial vapour pressures; this impairs the vacuum, and thus the *sublimation* proceeds more slowly. Further, the solvent-diluted pump-oil must be exchanged at frequent intervals.

For the reasons already mentioned, it is preferable to apply lyophilisation to salt-free solutions, or to select a salt concentration so low that the concentration attains the desired or permitted value only at the completion of the operation. Similarly, it is necessary to plan the pH change, particularly if the *buffer* mixture contains a volatile component.

The construction of the lyophilisation apparatus must also be carefully considered.

If the pressure in the system is less than 1.3 Pa (0.01 mmHg), the mass Q of water evaporated and condensed (related to unit surface area and unit time) may be calculated from the following formula, based on the kinetic gas theory:

$$Q = \frac{4}{3}\left(\frac{2\pi M}{RT}\right)^{1/2} \frac{r^2}{L}(P_a - P_b)$$

where M is the molecular weight of water, R *is* the universal gas constant, T is the absolute temperature, r *is* the radius of the cross-section of the vacuum line, L is the path-length from the substance to the pump, P_a is the effective water vapour pressure above the ice, and P_b is the partial water vapour pressure at the pump.

A whole series of various types of lyophilisation apparatus is manu-factured by the instrument industry; if a pump of suitable performance (0.01 Pa vacuum or better) is available, laboratory lyophilisation apparatus may also be constructed from *glassware*. Two-stage, gas-ballast valve, rotary oil-pumps are satisfactory for this purpose. A higher vacuum can be attained by the use of a diffusion-pump and an appropriate prevacuum-pump.

Preparation of Protein Solutions

Dissolution of tissue proteins

If the protein is to be obtained from animal or plant tissues, many factors must be taken into consideration, and it is necessary to reckon with the fact that the yield can scarcely be 100%. In this section we shall not touch on the problem of extraction of biologically active proteins; we shall attempt merely to describe those procedures by which protein solutions suitable for *analytical* purposes may be prepared without damage to the primary structure.

A decisive factor in the effectiveness of all protein-extraction procedures is attainment of the greatest possible degree of disintegration of the animal or plant tissue to be extracted. To break down the cell structure, apart from mechanical comminution, it is advantageous to employ disintegration by freezing or plasmolysis with *hypertonic* solutions, especially in the case of naturally occurring *moisture*-containing tissues.

The proteins display very varied solubilities. The scleroproteins do not dissolve in water, dilute salt solutions or other solvents in general. *Lipoproteins* can generally be dissolved out of the cell structure only after the extraction of the lipids. One group of plant proteins, the *prolamines* accumulating in the storage tissues of the endosperm of cereal seeds, dissolve only in 60-80% alcohol-water mixtures. The other proteins dissolve to various extents in water or in dilute aqueous salt solutions. The dissolution depends on the nature of the ions in the aqueous solution, their concentrations, the ionic strength, the pH value and the *temperature*.

At extreme temperature and pH the dissolution is accompanied by denaturing, and at the pH value corresponding to the isoelectric point the proteins are precipitated. Every procedure for obtaining proteins is at the same time a '*fractionation*'. Individual solvents may dissolve the different proteins to different extents, as the degree of dissolution depends not only on the composition of the solvent, but also on the structure of the protein. In this respect, the solubility depends primarily on the ratio of the polar hydrophilic and the non-polar *hydrophobic* groups, and on their arrangement, and consequently on the resulting dipole moment. Depending on the nature of the ions in the aqueous solution, very different quantities and types of proteins may be dissolved out of a given material, as the solubilities of proteins follow the general regularity known from colloid chemistry. The dissolution effect generally increases in accordance with the sequence of the *Hofmeister lyotropic* series. The peptizing effects of neutral salt and buffer solutions may be increased, for example, by the addition of anionic detergents. The sulphite-reduction of the disulphide bridges causes folding of the peptide chain and so increases the dissolution. The solubility of a protein is similarly increased by complex formation with the Cu^{2+} ion at pH > 8, but this is accompanied by a more profound structural transformation.

If a number of different solvents are used one after another to obtain proteins, then the resulting fraction distribution will depend on the sequence of solvents employed. For instance, if the proteins of cereal seeds are fractionated by the conventional Osborne method, the sequence of solvents is: water, dilute salt solution, 70% alcohol, dilute alkaline solution. If the extraction is begun with a basic solution, this will dissolve out practically all the proteins.

In their procedure for the extraction of maize and other plant seeds, Mertz and Bressani obtained good separation of protein fractions by dialysis after extraction with alkali and *cupric sulphate*. Kent-Jones and Amos reported a simple procedure for dissolving *albumin* and *globulin* out of cereal seeds with a 5% *potassium sulphate* solution; *methanolic alkaline* extraction was then performed, and the glutelin was precipitated by adjustment of the pH to 6.4, the gliadin remaining in solution.

The extraction of *albumin* and *globulin* from animal tissues is very difficult, for the tissues always contain ions, and these may be extracted together with the albumins, globulins and various *pseudoglobulins*. The individual fractions may be separated by precipitation, dialysis, *gel filtration* and other techniques. *Helander*, for example, extracts the *plasma protein* from muscle with *potassium phosphate* buffer at pH 7.3.

Pressing-out of tissue proteins

The soluble *protein* contents of plant or animal tissues may also be obtained by pressing. With laboratory-scale *hydraulic* presses, a pressure of 300-400 kg/cm may be achieved. The casing of the press is made of stainless steel, with run-off openings at the bottom. The sample thus obtained does not represent the total protein content of the material under examination, but only that part of the soluble protein that is not bound to structural elements and is obtainable by pressing at the given pressure.

Physical Seperation

Filtration and filters

In laboratory practice the most frequent means of separating solid particles from a *dispersion* in a liquid phase is filtration on fibrous or porous layers.

Filtration can be done by gravitation, suction or pressure. Various types of apparatus are used: filter *funnels*, *Buchner* funnels or sintered-disc filters and *pressure* filters.

The most important characteristics of filters are the retentivity and permeability (which determine the fineness of the particles which the filter is able to retain or let through), and the filtration rate. The filtration rate is generally taken as the volume of distilled water passed per unit area of filter in unit time under the suction pressure to be used.

Of the fibrous filters, filter papers are used most often in general practice. In the *protein-analysis* laboratory their use is mainly restricted to qualitative operation. Filter papers are made of short-fibre cotton, consisting of α-cellulose of about 95% purity. For special purposes, filters are also made from *glass*, *PVC*, *polyester* or *asbestos* fibre. The filtration characteristics can be described only qualitatively.

In protein analysis cellulose filter papers may be used for collection of the coarse dispersed fractions of suspensions, and of proteins coagulated with precipitating agents. For the first purpose, qualitative papers of high permeability are employed. Hardened papers have the advantages that they are resistant to acids and bases up to a given concentration limit, even at 20-50 °C, and their surfaces are smooth, so precipitates can readily be washed off them.

Schleicher and Schi ll produce a filter paper with low nitrogen content (N 2095), especially for protein *analysis* purposes. This is mainly used when, to avoid losses, the filter paper is mineralised together with a precipitate of low nitrogen content in a Kjeldahl *nitrogen* determination.

The advantage of filters prepared from *glass* and *plastic* fibres is that they are able to resist *corrosive* chemicals at even higher temperatures. Although these properties are rarely utilised in the protein-analysis laboratory, the filters have the added advantage that they do not swell or adsorb in aqueous medium. Glass-fibre paper may be used for prefiltration before use of ultrafilters. Very coarse dispersions are filtered on cheese-cloth or glass-fibre cloth in preparative procedures. Special impregnated filter papers are also utilised in certain cases. The *impregnating* agent may be an adsorbent, such as active carbon or kieselguhr. If these are used, it is also necessary to reckon with losses arising as a consequence of undesired adsorption. Active-carbon papers are suitable for the removal of dyes. Filter paper impregnated with kieselguhr, with a silica content of about 20%, is able to retain very finely dispersed particles of a *semicolloidal* nature. Filter papers treated with silicone may be employed to separate *aqueous* and *lipophilic* phases.

Table 2.5 : Pore sizes of sintered-glass filters (sintered-porcelain filters are similar).

Number	Jena filters μm	Pyrex filters μm
G-0	150-200	—
G-1	90-150	100-120
G-2	40-90	40-50
G-3	15-40	20-30
G-4	9-15	5-10
G-5	1-1.7	2

Porous filters are made of glass or porcelain. For downward suction filtration they are prepared in the form of filter *plates*, *crucibles* or *funnels*, and for upward filtration are in the form of immersed *plate filters*, *filter sticks* and *filter candles*. Buchner and Hartley funnels are used with filter papers; asbestos can be used in a Gooch crucible. Otherwise sintered-discs are used, made of glass or porcelain. Their *porosities* are well defined, and are denoted by the numbers of the filters.

Sintered-glass filters are made thicker or thinner, according to the diameter of the filter plate, so that they can withstand a pressure of 1 atm; filter plates of larger diameter are made slightly convex to increase the bearing strength. Filters 0, 1 and 2 may also be used for gravitational filtration of liquids of low *viscosity*. With the finer-pored filters, suction is applied with a *water-pump* or *rotary vacuum-pump*. A suction pressure of 2700-8000 Pa (20-60 mmHg) is generally sufficient for filtration at a satisfactory rate. The porosity of the filter must be chosen so that the diameter of the dispersed particles is somewhat larger than the maximum pore diameter of the filter, and in this way clogging of the filter plate can be avoided. It must be noted that the nominal pore size and flow rate constantly vary in the course of filtration, because the precipitate held on the filter diminishes the flow rate and makes the porosity finer. The G-5 filters are so fine that they may be used as filters for *bacteria*. This decelerating effect ofthe precipitate settled out on the filter may be decreased if auxiliary filtration materials are added to the suspension, though with protein solutions, adsorption must be reckoned with. Kieselguhr is the most satisfactory additive for biochemical work, e.g. '*Hyflo Super Cell*', which has a particle size of 5-25 μm. The use of *cellulose* is not advisable in *protein chemistry*. Glass filters may be sterilised, but they should not be cleaned with *chromic-sulphuric acid*, because chromium compounds may be adsorbed on the glass and disturb the biochemical analysis. *Perchloric-sulphuric* acid is a very good cleaning mixture, however. They may be dried or sterilised by slow heating up to 120-160°C, but a higher temperature or a sudden change in temperature may give rise to stresses in the filter, which lead to its cracking.

Ultrafiltration

Ultrafilters are based on permeable membranes, which separate dispersed ionic and molecular particles and the dispersing medium (water) from colloidal particles, and thus simultaneously concentrate the macromolecular solution. The wide choice and well-defined porosities of the industrially-produced membrane filters even allow a certain fractionation of the *macromolecules* according to molecular weight. However, the shapes of the molecules must also be taken into consideration: fibrillar macromolecules pass through an ultrafilter of given porosity in a different manner from globular molecules with the same molecular weight.

Depending on their structure, the filter devices may be either open systems suitable for continuous operation, or closed systems with intermittent operation. The through-flow of the liquid is assisted by suction or pressure. The small-scale laboratory ultrafilters are set up as filter funnels operated by vacuum suction. The intermittent ultrafilters, operating under pressure, are closed vessels in which the liquid to be filtered is placed and forced through under a pressure of $9.81 \times 10^4 - 98.1 \times 10^4$ Pa (1-10 atm) applied from a cylinder of compressed *nitrogen*. Very small membrane filters operating under *mechanical* pressure are also made in the form of an injection syringe. Continuous pressure filters, on the pattern of *industrial filter presses*, are also made on the laboratory scale. They have the advantage that the liquid to be filtered flows parallel to the filter surface and in this way the material sedimenting out does not form a layer obstructing filtration. The solution to be filtered may be transferred to smaller filters from a reservoir by means of *nitrogen* pressure. At lower and higher pressures it is customary to use a *peristaltic-tube pump* and a *displacement pump*, respectively.

The deposition of the coarse dispersed particles on the filter surface may be prevented by using fine *glass-fibre* prefilters.

Centrifugal ultrafiltration

In a special case of ultrafiltration under pressure, the ions and low molecular-weight substances, together with the dispersing solvent, water, are forced through the filter not by gas or mechanical pressure, but by an increased *gravitational* force. The method is therefore suitable for the separation of protein solutions from the ionic and molecular-weight dispersed particles, with simultaneous concentration. The '*Centriflo*' ultrafilter cones produced by Amicon N. V., The Hague, can be fitted into a special centrifuge tube. Swinging-head centrifuges are best used. With the aid of CF-50 ultrafilter cones at about 1000 g, 5 ml of 0.01% albumin solution can be concentrated to a volume of I ml. The CF-50 ultrafilter retains substances with a molecular weight of 5×10^4, but has no significant retentivity for molecular weights below 2×10^4.

In current laboratory operation, artificial membranes with well-defined *permeabilities* are used exclusively. The previously employed membranes of animal origin have lost their importance in laboratory practice. The gelatin filter, which is readily prepared in the gel state but easily damaged, is rarely used in the laboratory. The semipermeable filters employed are solid-structure membranes that may be stored in the dry state; they are produced industrially from various materials, with pore sizes corresponding to the particular purpose. However, since they can be prepared simply in the laboratory, collodion membranes are still used for dialysis.

The membranes may be made of cellulose nitrate, cellulose acetate, regenerated cellulose, *pol* (*vinyl chloride*) and *polyamide* (*nylon*). For purposes of dialysis they are prepared in the form of tubes and sheaths, and in the form of sheets for ultrafilters and electrodialysers.

Numerous descriptions are to be found in the literature with regard to the preparation of the collodion (*dinitrocellulose*) membranes employed in the laboratory. A *semipermeable* membrane may be prepared on some support. Filter paper is immersed in water, laid on a glass plate, and coated with a 4% collodion solution (in alcohol-ether, 1 : 7). The excess of solution is drained off, and after the *collodion* membrane has hardened (5-10 min) the filter is coated with a further layer. When this layer too has hardened, the filter paper is again immersed in water to dissolve the alcohol out of the layer. After about 30 min the filter is ready for use. In a similar manner a semipermeable membrane can also be layered onto an extractable sheath. A membrane can also be prepared without a supporting layer: a collodion solution (200 ml of 6% collodion solution diluted with a mixture of 200 ml of ether and 500 ml of alcohol) is poured onto a horizontal sheet of glass; after gelation has taken place the glass plate is immersed in water, and after a time the layer may be peeled off. A readily-separable collodion sheet filter is also obtained if collodion is poured onto the surface of mercury. The commercially-available membranes have the advantage that their pore sizes are well defined. The *permeabilities* of collodion membranes prepared in the laboratory, however, depend on the conditions of *gelation* (*solvent ratio*, temperature, etc.). Filters prepared from old collodion solutions have much larger pore sizes than those made from fresh ones. A collodion solution in *glacial acetic acid* may be aged by heat-treatment at 90-98 °C for 150 min, and thus filters with wider pores may be obtained.

Determination of the pore size is discussed in handbooks on practical *colloid* chemistry. It can be done by filtration of dye sols of known dispersities, or by measurement of the minimum pressure necessary to press air bubbles through.

The most important characteristic of commercially-available membrane filters is the pore diameter, which varies between 10 and 0.05 μm, depending on the material used and the manufacturing technology. Since the basis for separating the dispersed particles is the sieve effect, i.e. the difference between the dimensions of the dispersed phase and the pores, particles larger than the pores are generally deposited on the surface of the membrane, without any adsorption. The efficiency of filtration is independent of the operating pressure, the stepwise *gradient* and the *electrical* potential. These factors affect only the rate of dialysis or ultrafiltration. *Electron microscopy* shows that the pores occupy about 80% of the volume of a filter, their number being around 10'/cm.

Filters with the following pore sizes can be prepared from the materials listed:

cellulose nitrate	8-0.01 μm
cellulose acetate	0.5-0.2 μm
regenerated cellulose	0.6-0.2 μm
PVC	*ca.* 0.3 μm
polyamide (nylon)	*ca.* 0.8 μm

Table 3.6 : Comparison of various ultrafilter membranes.

Pore diameter, μm	Molecular weight excluded	Sartorius	Amicon	Millipore	Kalle
20—10	1×10^5	SM 11533	XM-100		
15—10	5×10^4	SM 12133	XM-50	PSDM	
5	3×10^4	SM 11539	PM-30	PSED	
5	1×10^4	SM 12136	PM-10		Dialysing
	1×10^4		UM-10		tube
	1×10^3		UM-2	PSAC	
	5×10^2		UM-05		

Cellulose-based filters can be sterilised at 130-180°C. PVC and *polyamide* cannot endure a higher temperature, but they are highly resistant to corrosive compounds. The thickness of membrane filters lies in the range 50-500 μm. The rate of through-flow of the pure solvent depends on the thickness. The through-flow rate is generally referred to that of water, which at a pressure difference of 93.3 kPa (700 mmHg) is 1 ml • min^{-1} • cm^{-2} at 25°C.

Depending on the thickness, the through-flow rates vary between 100 and 0.002 ml • min^{-1} • cm^{-2}. The Table elsewhere in this chapter lists the pore sizes and corresponding separating capacities of some of the membrane filters used in the *protein-analysis* laboratory.

Seperation by dialysis and electrodialysis

Dialysis

The essence of simple dialysis is the application of diffusion combined with the sieve effect for the separation of macromolecules from ions and compounds with low molecular weights. In its simplest form it requires only the dialysis tube. The liquid to be dialysed is poured into a membrane sheath (a *dialysis tube*), and this is immersed in pure water. The diffusion of the ions and small molecules into the water begins as a consequence of the concentration gradient. The rate of dialysis depends on the ratio of the volume of the solution to be dialysed and the surface area of the *dialysis* membrane, on the through-flow rate permitted by the *dialysis* membrane, on the magnitude of the concentration gradient between the internal and external solutions, and on the temperature. In accordance with *Fick's diffusion law*, the rate of dialysis is given by the following equation:

$$\frac{dS}{dt} = KA\frac{(C_k - C_b)}{dx}$$

where S is the amount of substance diffusing in time t through a membrane of surface area A, thickness d and diffusion constant K; C_k and C_b are the concentrations of the diffusing substance on the two sides of the membrane.

For a membrane of given quality, the rate of dialysis can be enhanced most simply by increasing the gradient of the external and internal concentrations. At *equilibrium* the dialysis will come to a stop. The gradient may be maintained by continuous exchange of the external solution, and by mixing, which prevents increase of the concentration at the interface. If ions are being removed, the gradient can also be maintained by adding ion-exchange resin to the external solution, or by circulating the solution through a mixed-bed ion-exchange resin column. The advantage of this is that the substance removed can be recovered if required. If it is desired to concentrate the *non-ionic*, *low molecular-weight* constituent, the Hahn apparatus is suitable. The dialysis sheath of this is placed in a closed system fitted with a *vacuum-distillation* condenser. The condensing vapour flows onto the *dialysis sheath* and is led by an overflow into the distillation flask, where the low molecular-weight component is concentrated. For maintenance of the concentration gradient, an economic method is recommended by *Hospelhorn*, who winds a thin cord spirally on the dialysis sheath suspended in a *cylindrical* vessel; the pure dialysing liquid flows throughout on this at a rate of 2 ml/min.

Dialysis may also be used instead of ultrafiltration to concentrate protein solutions by removal of the ions, low molecular-weight compounds and the dispersing medium (water) through the pores of the dialysis sheath under vacuum or pressure. Hence, the high molecular-weight solution is concentrated. A good description of the practical details is given by von Hofsten and Falkbring.

Electrodialysis

The final stage of dialysis, when the electrolyte concen-tration is very low, is extremely long drawn-out. Electrodialysis is mainly used to shorten this stage. A large proportion of the *electrolytes* is removed by common dialysis, and this is followed by *electrodialysis*. The efficiency of electrodialysis is increased by the simultaneous application of *ionophoresis*. The simplest type of electrodialyser is that with three cells. In this apparatus the solution to be dialysed is held between two membrane filters, in contact with the anode and cathode compartments; the wash-liquid, generally tap-water, flows into these two compartm-ents. The rate of dialysis can be regulated by adjusting the flow of wash-water, and also the potential. Depending on the electrolyte concentration, the potential must be regulated so that a power of *ca.* 1-2 J is available per cm^2 of the dialysis membrane. The detrimental effect of increased temperature may be diminished by increasing the flow of the wash-liquid, to act as a coolant. *Electrodialysis* may be disturbed by a charge on the membrane. *Synthetic membranes* in general have negative electric charge; they are therefore easily permeable to cations, but *hinder* the through-flow of anions as a result of the like charge. This effect can be eliminated by the use of a positively-charged membrane on the anode side, and a negatively-charged membrane on the cathode side. Breakdown of the ionic equilibrium as a result of the diffusion of H^+ and OH^- ions at different rates may give rise to a shift in the pH value in the middle (*dialysis*) cell, and consequently the protein solutions may undergo irreversible damage. By appropriate selection of the membranes, it may be arranged that the anions and cations depart in equivalent amounts. The colloidal particles accumulate and are possibly adsorbed on the surfaces of the polarised dialysis membranes, through which they are not capable of permeating. It is advisable to inhibit this process by stirring the solution to be dialysed, or by reversing the polarity of the supply potential for a short period from time to time, in this way changing the direction of dialysis. In the case of platinum or graphite electrodes this can be done without any problem, but a nickel *electrode* may only be connected as the cathode. The progress of the dialysis can be checked by conductivity measurement. It is also expedient to check the pH of the solution to be dialysed, with a glass-electrode *pH-meter*. Cells are often connected together in series to increase the dialysis surface area, and to improve the ratio of the surface area to volume; in these, washing and dialysis cells are connected alternately, separated by dialysis membranes. The wash-liquid flows from one washing cell to another in a *labyrinth* system. The dialysis cells are similarly interconnected.

A combination of *electrodialysis* and ion-exchange may be employed for the removal of salts from protein hydrolysates before paper or thin-layer *chromatography*. The anode and cathode cells of the three-cell electrodialyser are separated by anion-exchange and cation-exchange membranes, respectively, from the hydrolysate to be made salt-free. The ion-exchange membranes have extremely narrow pores, which practically let through only water. The ions diffusing towards the electrodes are exchanged on the ion-exchange membranes. The *anode* compartment is filled with 0.2 *M sodium hydroxide*, and the *cathode* compartment with 0.2 *M* sulphuric acid. The cation-exchange membrane lets through only the cations, with the exchange of H^+ ions, while on the anion-exchange membrane OH^- ions are exchanged for the anions of the solution to be made salt-free. In the low-volume middle cell, the solution from which the salt is to be removed is kept in motion by the passage of bubbles through it. In an apparatus with favourable geometry, the ion-exchange proceeds very rapidly at a potential of 6 V, as is readily seen from the fall in the current passing through the cell, as a consequence of the increase in resistance. The salts can be removed from the protein *hydrolysate* in 6- 10 min. A certain degree of loss of basic amino-acids may occur; this may be avoided if the starting pH of the solution to be made salt-free is adjusted to 8.0 with 0.05 *M potassium carbonate*.

Centrifugation and Centrifuges

One of the general laboratory methods used for separation of the dispersed particles of emulsions and *suspensions* is centrifugation. The development of preparative and *analytical* methods in protein chemistry is related in many respects to the development of laboratory *centrifuges*.

The essence of centrifugation is the faster sedimentation of the dispersed particles in a *gravitational* field that is increased as a consequence of rotary motion.

The Stokes sedimentation law applies to centrifugation:

$$V = \frac{2}{g}\frac{\rho^2(d - d_f) - g}{e}$$

where V is the rate of sedimentation, p is the density of the spherical particles, d is the density of the dispersed particles, d_f is the density of the dispersing medium, g is the gravitational acceleration, and e *is* the internal frictional coefficient of the medium.

A force $Mr\omega^2$ rad/s^2 (radial acceleration) acts on every particle of the dispersed system rotated in the centrifuge, where M is the mass of the particle (g), r is the rotation radius (cm), and ω is the angular velocity (rad/s).

In practice the relative centrifugal force is used, or, in another notation, the g value of the centrifugation. This may be calculated simply from the formula:

$$g = \frac{f^2 r}{89\,500}$$

where f is the rotation rate of the centrifuge in rpm, and r is the rotation radius measured in cm from the axis.

The time necessary for the sedimentation of some fraction with a definite degree of dispersity can be calculated approximately from the formula:

$$t = \frac{D - L}{D + L} \cdot \frac{N}{a^2(d - d_f)f^2}$$

where D *is* the length of the rotation radius from the axis to the meniscus of the suspension, N is the viscosity of the dispersing medium in poise (g·cm^{-1}·s^{-1}), d *is* the density of the material of the dispersed particles, d_f is the density of the dispersing medium, a is the average diameter of the dispersed particles, and f is the rotation rate of the centrifuge in rpm.

In preparative and analytical *ultracentrifugation*, attention must be paid to the following factors determining the sedimentation time.

Non-spherical particles sediment more slowly than spherical particles with the same mass and density, for which the Stokes formula holds.

The rate of sedimentation depends on the temperature. If the rotation rate and g value are increased, the sedimentation rate is decreased by the heat produced by friction. Accordingly, the cooling of the centrifuge must be adjusted to counterbalance the heating.

In description of laboratory methods, the conditions of centrifugation may be characterised unambiguously by giving the centrifugation time, the g value and the temperature.

For comparison of results obtained under different experimental conditions, the sedimentation rate of a definite fraction is reduced to a value of g=1 and a temperature of 20°C. This value is known as the sedimentation constant; it is denoted by the symbol S_{20}, and has units of cm/s.

Centrifuges may be classified in terms of rotation rate and type of rotor. If the rotation rate is used as the criterion, we may distinguish:

low rotation-rate centrifuges:	0-5000 rpm
high rotation-rate centrifuges:	0-20000 rpm
ultracentrifuges:	0—60000 rpm

In terms of rotor construction, the classification is: angular centrifuges, swinging-head centrifuges, through-flow separators.

The *rotor* of an angular centrifuge is a truncated *conical* body made of a high-strength light metal or plastic; in this holes are drilled radially and *symmetrically*, at an angle of 15-60° to the rotation axis, for insertion of the centrifuge tubes. The g value varies along the length of the centrifuge tube, depending on the distance from the rotation axis. The maximum path-length for the dispersed substance to be sedimented is the *horizontal* diameter of the centrifuge tube, which depends on the angle of the drilling.

In swinging-head centrifuges, the centrifuge tubes are inserted into containers free to move in the forked supports from which they are suspended and which are arranged radially symmetrically with respect to the axle. As a consequence of the centrifugal force, the containers are horizontal during operation. The g value varies along the tube, depending on the distance from the rotation axis; it is customary to quote the value at the bottom of the centrifuge tube. The maximum sedimentation distance of the dispersed particles is the height of the suspension in the centrifuge tube.

The rotor of a through-flow *centrifuge* is a tube or drum, and the suspension is led into it from below through a hole in the axle. The separated fractions are led off at the upper rim of the cylinder, through ports situated at various distances from the rotation axis. The efficiency of sedimentation depends on the inflow rate of the suspension as well as on the technical parameters of the centrifuge.

Besides these preparative-type rotors, analytical type rotors are produced for *ultracentrifuges*; these permit continuous monitoring of the sedimentation by *refractometric* or *interferometric* means, on the basis of the principle elaborated by Svedberg and Pedersen and subsequently further developed by numerous instrument companies. In recent years the instrument companies have constructed a very varied series of rotors for purposes of various biochemical preparative techniques, and particularly for the isolation of cell fractions. These permit *centrifugation* under conditions where the dispersing medium, e.g. a sugar solution, forms a density gradient in the drum rotor (preparative zonal centrifugation) or in a centrifuge tube specially constructed for this purpose (Strohmaier fractionation tube.

The technique of centrifugation may be combined with other separath procedures, e.g. filtration. For this purpose dismountable centrifuge tubes fitted with a special filter insert are produced.

Normal centrifuge tubes are made of glass or plastic (*polyethylene*, *polypropylene* or possibly *cellulose acetate*). It is not recommended that glass centrifuge tubes should be used at a rotation rate higher than 3000-4000 rpm.

Centrifuges with relatively low rotation rates are driven by a universal *electric motor*, with a variable resistance in series to ensure continuously variable rate regulation.

In centrifuges with high rotation rates, the effect of heat is diminished by using an oil-pump built into the centrifuge to keep the working area of the rotor under vacuum (the working area is enclosed by a strong protective sheath); constant low temperature is ensured by a cooling spiral surrounding the working area.

GENERAL CHARACTERISATION

From the classification of the substances examined in protein analysis it may he seen that their detailed investigation differs to various extents, depending in part on their origin and in part on their use. However, because of their use, the complete analysis of certain protein-containing materials is pointless, e.g. in the routine quality-control of food and feed proteins it is quite sufficient to measure those properties which show the nutritional value, knowledge of which promotes their optimum utilisation. These properties are the *moisture*, *ash* and *protein* contents. The methods for determination of the moisture and ash do not display great variety or involve any difficulty. The protein-determination methods are much more varied.

Lewis has described a method for characterisation of the variation in the sensitivity of some commonly used protein assay procedures.

Dry Matter

The analytical results are generally calculated on the dry weight. The basis of comparison of agricultural and industrial raw materials is not always the absolute dry weight, but that measured in accordance with conventions and standards; it is usually the moisture in the substance in *equilibrium* with the partial water vapour pressure of the atmosphere after a protracted period at room temperature in air with an average relative *humidity* of around 60%. For *commercial* goods, it is *customary* to use as the reference basis that dry matter or moisture content with which the goods can be stored without deterioration. In the *majority* of cases this value agrees with the 'air-dried' weight discussed above.

In naturally-occurring substances the moisture may be present as free water or intramolecular water. The most frequent method of determining the dry matter is drying at constant temperature. If the loss of water at constant temperature is examined as a function of time, a *characteristic* drying curve is obtained, in which there are three stages. In the first stage the water content decreases quickly as a result of the high evaporation rate from the surface and outer layers. In the second stage the rate of loss of water is determined by the increasing diffusional resistance of the drying substance. In the third stage the water-binding capacity of the third material is important, and the rate of loss decreases greatly. When we speak of drying at constant temperature to constant weight we are concerned with the attainment of this third stage. It must be stressed that, in terms of *laboratory* operations, determination of the dry matter by drying must be distinguished from drying of the substance. The drying employed for the determination, e.g. in the case of protein-containing materials, may lead to irreversible denaturing. In contrast, the drying of a substance in preparation for other examinations must be done in such a way that denaturing does not occur.

The main laboratory methods used for determination of the dry matter are classified into the following types:

(a) drying at atmospheric pressure and 100—105 °C or higher;

(b) drying in vacuum at below 100'C;

(c) drying by distillation of the water;

(d) determination by chemical means (e.g. Karl Fischer titration).

A 2-5 g sample of air-dried finely-ground material may be dried practically to constant weight by drying for 3 h at 105°C. With drying at 130 °C, weight constancy may be achieved within 1 h. A sample of a few *g* may be dried to weight constancy within 8-10 h in a vacuum drying oven at 70°C and 6.67 kPa (50 mmHg). The moisture may be determined by distillation with *toluene* or *xylene* in a *Dean* and *Stark* apparatus. The sample taken should yield 2-5 ml of water. The solvent is added, and distillation conducted so that 2-4 drops condense per second. The distillation is continued for about 1 h. The volume of water collected below the solvent layer in the calibrated receiver tube is measured. The same method is used for determining water in oil.

The most popular chemical determination is the Karl Fischer *titration* in one variant or another. The essence of the method is the reaction:

$$I_2 + SO_2 + 2H_2O = 2HI + H_2SO_4$$

When the concentration of *sulphuric acid* formed exceeds 0.05%, the reverse reaction, the oxidation of *hydrogen iodide*, begins:

$$H_2SO_4 + 2HI = I_2 + H_2SO_3 + H_2O$$

Naturally, the method works only in *non-aqueous* medium. The essence of the technique developed by Fischer is that the stabilities of the reagents are enhanced with the aid of methanol and pyridine, which form addition compounds with iodine and *sulphur dioxide*, and the reaction is driven to completion. The reaction is assumed to take place as indicated by the equation:

$$I_2 + 2H_2O + py_2 \cdot SO_2 + 2\ py \rightarrow py_2 \cdot H_2SO_4 + 2\ py \cdot HI$$

The visual end-point is unsatisfactory and in practice *electrometric* methods are used, with *potentiometric* or dead-stop end-point detection.

Ash Content

In the case of plant materials, a 2-3 g air-dried, finely-powdered sample is taken in a *platinum*, *porcelain* or *quartz crucible* and *ashed* initially over a low flame; as soon as the burning is over, ignition is continued over a strong flame until the ash is light in colour. In an ashing-oven, the maximum temperature used is 750°C to avoid loss of alkali metals. Sometimes a temperature of 525-550°C is recommended, to prevent loss of phosphorus. In plant raw materials the sand content is determined separately and the ash value corrected for any sand contaminating the outside of the plant material. In the case of foodstuffs, conserves, and especially meat products, the ash value must be corrected by subtraction of the *sodium chloride* content.

Fat Content

The fraction extractable from materials of plant and animal origin with fat-dissolving agents is known as *crude fat*. Besides the *triglycerides*, this crude fat contains very many other substances: *fatty acids*, *phosphatides*, *dyes*, *alkaloids*, etc. The methods of determining these do not belong n the present topic.

The determination of the *crude fat* and the removal of fat from the sample are essentially the same process. In work on an analytical scale or for small-scale preparative purposes, it is done in a Soxhlet extractor. Peroxide-free diethyl ether or petroleum ether (40-60 °C) is used as the solvent. An air-dried, finely-ground sample is extracted for 5-6 hr at the rate of 8-10 *syphon cycles* per hour. An efficient condenser is needed, to give perfect reflux. The fat-containing solvent must be distilled off carefully, and the crude fat dried at 95 ± 2°C.

If the aim is merely the removal of fat from the sample, and no effort need be made to recover quantitatively either the fat-free sample or the crude fat, then the fat may be removed by homogenizing the air-dried material with a non-inflammable solvent (carbon tetrachloride, chloroform) in a blender, filtering off on a glass filter, and washing the residue several times with pure solvent.

Kjeldahl Method

Most indirect protein-determination procedures are based on determination of the nitrogen content, because the elemental compositions of the different proteins are very similar, independent of the nature and origin of the protein.

Naturally, the nitrogen content may be used only to determine pure proteins or the protein content of a material that includes no other nitrogen-containing matter than protein. In *agricultural* and *food analysis*, the total protein content of the sample is determined as 'crude protein', by multiplying the *nitrogen* content by 6.25, a factor corresponding to an average *protein nitrogen* content of 16%.

In practical analysis this yields accurate enough results, but if an exact result is required, the factor employed must correspond to the *nitrogen content* of the individual protein. The nitrogen contents and *protein conversion* factors of some proteins and protein-containing substances.

The Dumas method for nitrogen determination was published in 1831 and was the only reliable method until the Kjeldahl method was described in 1883. The latter, with its numerous modifications, has remained the basis of *biochemical nitrogen* determination ever since; this has had a *tremendous* effect on the development of biochemical and applied biochemical analysis.

Table 3.7 : Nitrogen contents of proteins, and factors for calculation of the protein from the nitrogen content.

Protein	Origin	Protein nitrogen content, %	Factor
		Proteins	
Plasma albumin	human	15.95	6.27
Plasma albumin	bovine	16.07	6.22
Ovalbumin	hen-egg	15.76	6.34
Globulin	human plasma	16.03	6.24
Lactoglobulin	cow's milk	15.60	6.41
Casein	cow's milk	15.63	6.39
Haemoglobin	equine blood	16.80	5.95
Fibrin	human plasma	16.90	5.91
Fibroin	pure silk	18.70	5.35
Keratin	feathers	15.00	6.66
Keratin	pig-bristle	16.60	6.02
Keratin	wool	16.30	6.13
Collagen	bovine skin	18.60	5.37
Gelatin	collagen	18.00	5.55
Histone	calf-liver	18.00	5.55
Histone	calf-thymus	18.20	5.49
Protamine	bird sperm	24.40	4.10
Protamine	salmon roe	30.70	3.25
Gliadin	prolamine from wheat	17.66	5.66
Hordein	prolamine from barley	17.21	5.81
Zein	prolamine from maize	16.20	6.17
Legumin	pea	16.04	5.54
Glycinin	soya-bean	17.30	5.78
Edestin	hemp-seed	18.55	5.39
	Naturally-occurring materials containing protein		
Cereal seeds			
barley		17.15	5.83
wheat		17.15	5.83
wheat-endosperm		17.54	5.70
wheat bran		15.84	6.31
wheat germ		17.23	5.80

Table 3.7 (Contd.)

Table 3.7 (Contd.)

Protein	Origin	Protein nitrogen content, %	Factor
maize		16.00	6.25
rye		17.15	5.83
oat		17.15	5.83
rice		16.80	5.95
Leguminous plant seeds			
bean, pea, lentil		16.00	6.25
Oleaginous plant seeds			
sunflower		18.86	5.30
soya-bean		17.51	5.71
coconut		18.86	5.30
cotton-seed		18.86	5.30
Materials of animal origin			
egg		16.00	6.25
meat		16.00	6.25
milk		15.67	6.38

Kjeldahl decomposition

Much of the research on the Kjeldahl method has been devoted to perfection of the decomposition step. The chemistry of the decomposition with *sulphuric acid* is relatively simple. Concentrated *sulphuric acid* is an extremely effective *dehydrating* agent: at a given critical temperature it extracts the elements of water from oxygen-containing *organic matter*, and liberates carbon. At the decomposition temperature sulphuric acid also acts as an oxidant and oxidises the liberated carbon to *carbon dioxide*.

In the case of saturated unbranched-chain primary amines, the reaction can be generalised as follows:

$$C_nH_{2n+1}NH_2 + mH_2SO_4 \rightarrow nCO_2 + (m-1)SO_2 + 4n\ H_2O + NH_4HSO_4$$

where n is the number of carbon atoms, and m is the sum of the numbers of carbon and hydrogen atoms in the alkyl radical. For amides:

$$C_nH_{2n-1}CONH_2 + (m+1)H_2SO_4 \rightarrow (n+1)CO_2 + (m-1)SO_2 + (4n+1)H_2O + NH_4HSO_4$$

For amino-acids:

$$NH_2(CH_2)_nCOOH + (m+1)H_2SO_4 \rightarrow (n+1)CO_2 + mSO_2 + 4nH_2O + NH_4HSO_4$$

The work by *Carpiaux* indicates that 1 g of protein consumes *ca.* 9.0 g of *sulphuric acid.*

The Kjeldahl decomposition is generally carried out in long-necked, pear-shaped Kjeldahl flasks. The long neck of the flask serves to condense the water vapour and to prevent too rapid evaporation of the decomposed liquid.

Several possibilities are available for acceleration and completion of the decomposition, such as raising the temperature, adding *oxidants*, adding *reductants*, using *catalysts*.

Table 3.8 : Effect of K_2SO_4 in elevating the boiling point of sulphuric acid.

K_2SO_4/H_2SO_4, g/ml	Boiling-point, °C
	0338
0.33	349
0.625	363
1.0	383
1.3	397
2.0	450

Elevation of the decomposition temperature.

On heating, pure sulphuric acid boils and decomposes at 338°C. By the addition of salts the boiling-point can be increased and the decomposition time for organic compounds shortened, and in this way the nitrogen content of certain substances that are otherwise difficult to break down can be converted *quantitatively* into *ammonia*. The salt generally used is *ammonium sulphate*. McKenzie and Wallace have shown that the boiling point of concentrated *sulphuric acid* varies with the *potassium sulphate* concentration.

However, the addition of *potassium sulphate* not only raises the boiling point, but also consumes *sulphuric acid* by formation of *potassium hydrogen sulphate*:

$$K_2SO_4 + H_2SO_4 \rightarrow 2\ KHSO_4$$

On the basis of this equation, 1 g of potassium sulphate reacts with 0.56 g of *sulphuric acid* and this must be allowed for in the volume of *sulphuric acid* added.

Numerous other salts have been tried, including *sodium sulphate*, *lithium sulphate* and *borax*, but these do not display any particular advantage. The same may be said for *phosphates*, which consume even more *sulphuric acid* than *potassium sulphate* does, with the formation of acid salts.

The addition of salts is limited by their solubility in *sulphuric acid*. The decomposition temperature can also be raised by addition of a *non-volatile acid*, such as *orthophosphoric acid*, but there is often a decrease in *oxidising power* and hence in the rate of oxidation. Experience has shown that it is not expedient for the proportion of phosphoric acid in the decomposition mixture to exceed 40%, and in terms of temperature this mixture has no advantages over a *sulphuric acid-potassium sulphate* solution. It is much better to add *potassium sulphate* to a *sulphuric acid-phosphoric acid* mixture, the best composition being 0.45 g of the salt per ml of 80% sulphuric acid-20% *phosphoric acid* mixture. The initial boiling point is 354°C, and progressively rises during the decomposition. At the boiling point the rate of decomposition of the sulphuric acid is 0.081 ml/min, and during a 90 min decomposition the acid ratio changes from 80:20 to 76:24 and the boiling point to 364°C.

Another possibility of raising the temperature is decomposition in a sealed glass tube, because of the increase in pressure. This technique is difficult and dangerous, so is only worthwhile if accompanied by a substantial advantage. It does have the advantage of needing no catalyst. Even a very small amount of *ammonia* can be determined by *alkaline* distillation from the decomposition solution and *colourimetric* determination, e.g. by the *Nessler method*, but catalysts may interfere in the *Nessler reaction*, which is why decomposition in a sealed tube without a catalyst is advantageous. With this technique, the decomposition temperature may reach 490°C, even when potassium sulphate is not added, but care must be taken that it does exceed 500°C, for in that case nitrogen might be formed, which would be lost when the tube was opened. Only dry samples may be decomposed in a sealed tube; if the material is in the dissolved state, it must be dried by *vacuum-evaporation* or *lyophilisation*, most simply in the decomposition tube itself.

Addition of oxidants to accelerate the decomposition

In his original method, Kjeldahl used powdered *potassium permanganate* in the final stage of the decomposition to improve the oxidation of the organic matter. For a long time the necessity of adding *permanganate* was debated, on the basis of contradictory analytical findings, but finally it was considered to be unnecessary and was discarded. *Numerous* publications have appeared with regard to the application of other oxidants, mainly ***hydrogen peroxide***, *perchloric acid*, *persulphates* and *chromic acid*. The question arises as to the value of the presence of an oxidant in the various stages of decomposition. It may be assumed that at the stage of *dehydration* and *carbonisation* the nitrogen is still not completely reduced, and thus an oxidant may inhibit the reduction. Nevertheless, efforts to shorten the decomposition time make the use of oxidants necessary. With naturally-occurring materials, which do not contain a large quantity of oxidised nitrogen, it does not prove disadvantageous, particularly when hydrogen peroxide is used. The method of Kleeman, for example, gives very fast decomposition, and also diminishes fume formation. The procedure is as follows: about 1-5 g of fresh plant or animal material is weighed into a Kjeldahl flask and, with shaking, 25 ml of 30% hydrogen peroxide are added, followed by 40 ml of concentrated sulphuric acid. It is found that about 80% of the total nitrogen is reduced very rapidly to NH_3. If 15 g of potassium sulphate are also added, boiling for 10-15 min is sufficient for quantitative reduction of the nitrogen, but Kleeman recommends 45 min. In control experiments with known amounts of $(NH_4)_2SO_4$, there was no nitrogen loss under similarly intensive oxidising conditions.

A number of modifications of the method are known. Heuss added 15 ml of 30% hydrogen peroxide, 20 ml of concentrated sulphuric acid and 7-8 g of potassium sulphate to 1.75 g of dry material; complete decom-position can be attained within 30-45 min. Riehm used a catalyst as well, adding 0.5 g of copper and 0.8 g of mercuric sulphate. Koch and McMeekin use hydrogen peroxide in a micromethod. In the *carbonisation* stage the heating is interrupted, 1-5 drops of ***hydrogen peroxide*** are added, and heating is resumed until white fumes again appear; this operation is repeated several times. Moore adds hydrogen peroxide to the decomposed material after mild heating for 5 min.

Perchloric acid can be used in a fairly similar way. Mears and Hussey add 1 g of copper sulphate and 2 ml of concentrated perchloric acid to 25 ml of concentrated *sulphuric acid*; the solution becomes clear within 3-7 min, and must then be boiled for at least a further 15 min.

A number of methods are available in which persulphates are used. Willard and Cake employ persulphate in the determination of the nitrogen contents of *flour*, *albumin*, *gelatin* and *peptone*. The sample is heated with 15 ml of concentrated sulphuric acid until fuming ceases. The flask is cooled, a quantity of potassium persulphate weighing ten times as much as the sample is added, the solution is heated gently until it becomes clear, and then more strongly for 5 min to decompose the excess of persulphate. Too high an excess of *persulphate* may cause loss of *nitrogen*.

Swaminathan and Sud proposed a procedure for rapid determination of total nitrogen which consists of weighing a 0.5-5 g sample (fruits and vegetables) into a 500 ml Erlenmeyer flask and adding to it 20 ml of 50% solution of chromic acid followed by 25 ml of concentrated sulphuric acid with swirling. After 30 minutes the contents of the flask are diluted with water, transferred to a 100 ml volumetric flask, cooled and made up to the mark. A 5 ml aliquot of the digest is pipetted into a beaker, a trace of potassium persulphate is added and boiled for a few seconds to oxidise the Cr^{3+} present to Cr^{6+}. About 20% of the nitrogen was found to be bound irretrievably as *chromamine* complex. *Persulphate oxidation* of the Cr^{3+} was found to be the only method whereby all the nitrogen could be recovered in the distillate quantitatively.

Application of reductants

In studies of *nitrogen metabolism* it may be necessary to determine the total nitrogen contents in plant materials containing nitrate. In this case the nitrate nitrogen can only be converted into *ammonium* ions in the Kjeldahl decomposition if a reductant is added. Reduction is also necessary in the determination of the nitrogen of *azo*, *nitro*, *oxime*, *isoxazole*, *hydrazine* and *hydrazone* groups. The reduction may be done either prior to or simultaneously with the sulphuric acid decomposition. *Metallic zinc*, metallic copper, titanous chloride, sodium dithionite or benzoic acid may be used for the preliminary reduction. According to Jodlbauer, the reduction of nitrate is more effective if phenol is used instead of benzoic acid. In German books on soil analysis, the Jodlbauer procedure is applied in general for determining the total nitrogen contents of substances containing nitrate. In the AOAC methods book, *salicylic acid* and *metallic zinc* are recommended for enhancement of the reduction.

Catalysts

During the long history of the Kjeldahl method, the catalytic effects of various metals, metal oxides and a whole series of salts have been investigated with a view to shortening the decomposition time. Osborn and Wilkie found that the catalytic efficiency of the metals decreases in the sequence: Hg, Se, Te, Ti, Mo, Fe, Cu, V,

W, Ag. However, numerous other examinations have proved Hg, Cu and Se to be the most effective. Baker has shown that catalyst mixtures do not have demonstrable advantages over the individual catalysts, but in practice the use of various *catalyst* mixtures is generally *widespread.*

The destructive decomposition passes through various stages: the *dehydration*, the *carbonisation* and the clearing of the solution. The transformation of nitrogen into *ammonia* is typically exponential; the conversion is rapid at first, but becomes increasingly slower.

The stage at which the transformation curve is almost parallel to the time axis and the amount of nitrogen transformed approaches the absolute nitrogen value *asymptotically*, is termed after-boiling. According to Beet, it may be anything from 0 to 235 h. Kitto found that the time required for the after-boiling of wheat-flour and milk is 1.75 times the time necessary for clarification of the solution. In the case of substances which are difficult to decompose (*pyridine*, *nicotinic acid*), Shirley and Becker (using the AOAC method) considered a boiling period of 3-4 h after the clarification to be necessary in order to attain quantitative conservation. The *potassium sulphate* (added to raise the boiling point) and the catalysts are also of very great importance as regards the duration of the after-boiling period. The amount of acid used for the decomposition must be chosen so that, even after the after-boiling stage, sufficient remains to bind the ammonia formed. A loss will occur if, as a result of concentration of the *potassium bisulphate* and the *catalyst*, a solid phase is present when the decomposition is completed. Self recommends that at the end of the decomposition at least 15 g of sulphuric acid should remain free in the solution from the 30 ml initially taken. Before the investigation of large series of samples of similar nature, even in the case of standard methods, it is advisable to carry out a few preliminary experiments to establish the optimum after-boiling time necessary; in this way, the *analytical* error may be decreased, and unnecessary time losses may be avoided.

Determination of the ammonia

The second stage in the Kjeldahl method is determination of the *ammonia* formed. This can be done *titrimetrically* (with or without distillation) or *spectrophotometrically*.

Two distillation methods are available: direct *distillation* and *steam distillation*. Separation of ammonia by diffusion may also be regarded as a distillation process.

In *macrodeterminations*, direct distillation is generally used (the Kjeldahl flask itself being used as the distillation flask), and the *ammonia* is liberated with concentrated alkali. An alternative is to dilute the decomposition solution to known volume and use an aliquot in a microdistillation. Mercury forms a complex with ammonia, so if a *mercury catalyst* is used, the mercury must be removed before addition of the alkali, or a low result may be obtained. *Potassium sulphide* or *sodium thiosulphate* is used to precipitate the *mercury*. About 0.2 g of *thiosulphate* is required for of 1 g of mercury. Salm and Prager showed that, even in the presence of a *mercury catalyst*, loss of *ammonia* may be avoided by adding finely-divided zinc powder to the alkaline solution to be distilled. The zinc reduces the, mercury salts to -the metal. Another disadvantage of the mercury catalyst is that in micro and *semimicro* methods the precipitate of HgS may block the steam duct of the apparatus. By direct distillation with moderate heating, the *ammonia* can be distilled off quantitatively in 5-20 min, depending on the size of the apparatus. In the steam-distillation apparatus of Parnas-Wagner, Markham, Biichi or similar types for the *micro* and *semimicro* determination a maximum of 10 min is needed for the quantitative distillation of the ammonia.

In the classical Kjeldahl method, the ammonia is absorbed in sulphuric acid of known volume and normality, and the excess of acid is back-titrated iodometrically:

$$5KI + KIO_3 + 3H_2SO_4 = 3I_2 + 3K_2SO_4 + 3H_2O$$

$$I_2 + 2Na_2S_2O_3 = 2NaI + Na_2S_4O_6$$

A simpler method is absorption of the *ammonia* in acid of known volume and normality, and *back-titration* of the excess of acid with standard alkali.

Winkler recommends absorption of the ammonia in 4% boric acid solution; this has a pH of 4.7, which changes to 8.6 on about 20% neutralisation with *ammonia*. The *ammonia* thus absorbed in the boric acid solution can then be titrated with standard acid. Various indicators and indicator mixtures are used. Winkler suggested *Methyl Orange*, *Methyl Red* or *Congo Red*, and *Scales* and *Harrison Bromophenol Blue*. The equivalence point at pH 4.7 is particularly well indicated by the *Tashiro* indicator: a 1 : 1 mixture of 0.2% Methyl Red and 0.1% *Methylene Blue* solutions. In pure boric acid this mixture is purple, but at pH above 4.7 it changes to apple-green. Reith and Klazinga use a mixture of 1 volume of 1 % Methyl Red solution, 3 volumes of 1 % *Bromocresol Green* solution and 4 volumes of distilled water. The colour of this indicator mixture changes from *bluish-grey* to *violet*. The end-point may also be determined potentiometrically.

The thermal stability of *ammonium borate* has been studied. There is no ammonia loss from the ammonium borate solution during heating for 30 min at 50°C; at 60°C and 100°C, however, there are losses of 1.3% and 62.3%, respectively.

The diffusion method is suitable for the determination of very small amounts of *ammonia* (0.1-100 μg). The procedure was developed in a form suitable for routine work by Conway and Byrne. With the use of Conway dishes, this amount of ammonia may be absorbed in about 3 h at 20 °C. The method is rarely used in protein analysis, but may be employed very advantageously for ammonia determination in the investigation of *microbial nitrogen* metabolism.

A very rarely used method is formaldehyde titration:

$$2(NH_4)_2SO_4 + 6HCHO = (CH_2)_6N_4 + 2H_2SO_4 + 6H_2O$$

An aliquot of the diluted decomposition solution is neutralised in the presence of Methyl Red as indicator, and then, after the addition of *formaldehyde*, the *titration* is continued to a phenolphthalein end-point. The difference between the two *titration* volumes is equivalent to the *ammonia*.

In another method Willard and Cake oxidised the ammonia with *hypobromite*, and determined the excess of *hypobromite* iodometrically.

The *spectrophotometric* methods are mainly used for direct determination of the ammonia in the decomposition solution, without preliminary distillation, but are also more advantageous than titration for the determination of *ultramicro* quantities in the distillation method. If the ammonia is absorbed in boric acid, for instance, and 0.01 N acid is used for the titration, the end-point can be determined with an error of ca. 0.02 ml with the Tashiro indicator under ideal conditions, which corresponds to 2.8 μg of nitrogen. In contrast, with the high-sensitivity *ninhydrin* reaction it is possible to measure 0.007 μg of nitrogen in a final volume of 1 ml. However, there are many interferences in the colour-reactions of ammonia, for example, by the catalysts used in the Kjeldahl method, and this must be taken into account in selection of the method.

The colour-reactions of ammonia with the Nessler reagent, indophenol and *ninhydrin* are generally used for the determination.

The Nessler reagent is an alkaline solution of potassium tetraiodomercurate(II), from which a *cyclic* complex is formed by the action of ammonia:

$$2K_2HgI_4 + 3KOH + NH_3 \rightarrow O\begin{matrix}\diagup Hg \diagdown \\ \\ \diagdown Hg \diagup\end{matrix}N^+H_2—I^- + 2H_2O + 7KI$$

In the presence of a little ammonia, a yellow colour is obtained; with somewhat more ammonia, the colour is orange; and with much ammonia a *brownish-red* precipitate is formed. The *sulphuric acid* solution from the decomposition must be diluted and *neutralised* before addition of the reagents.

According to Roth, the Nessler method is advantageous if the quantity of nitrogen to be determined is in the order of 10 μg. One of the drawbacks of the methods is that the Nessler reagent cannot be stored for a prolonged period. Yuen and Pollard made a study of the factors that govern the reaction. They found that the colour intensity produced depends on time, *temperature*, the amount of reagent, the nature and concentration of other ions, and pH. Hence identical experimental conditions must be used and a standard run simultaneously. Various wavelengths are recommended, but the measurement is usually made at 430-450 nm.

Even under optimum conditions, a turbidity may develop. This may be prevented with stabilisers such as *tartrate*, *cyanide*, and *gum arabic*.

It is advisable to choose decomposition conditions such that the concentration of the *catalysts* and *potassium sulphate* will not be too high in the final solution taken for the colour reaction.

The *indophenol* reaction is widely used and is also the basis of the *automatic analyser* methods. Phenol, salicylic acid and even thymol are used, with sodium or *potassium hypochlorite* or *hypobromite* as oxidant.

As reported by *Berthelot* as long ago as 1859, the essence of the reaction is that in the presence of *hypobromite p-nitrosophenol* is formed from *phenol* and *ammonia* and in *alkaline* medium is then converted into a light-blue indophenol *alkali metal* salt. Riley demonstrated that temperature and time are the critical factors. The reaction is catalysed by the Mn^{2+} ion. Exley developed the phenol-hypochlorite method into an ultramicro procedure, and successfully utilised it for the analysis of *insulin*, *casein*, *egg-white*, *bovine* and *human sera*, etc.

Mann's method is suitable for the determination of 1-15 μg of nitrogen. The final dilution must yield an ammonia concentration corresponding to 0.1-1.5 μg of nitrogen per ml. At this concentration, with measurement at 630 rim, the absorbance obeys the *Beer-Lambert* law, by an adaptation of the method of *Ogg*. It is important that all experimental conditions be kept strictly constant. The colour intensity depends on pH, temperature and the reagent concentrations, but especially on the *hypo-chlorite* concentration, and this reagent should be checked by *iodometric titration* before every test series. *Glebko et al.* consider the *thymol-hypobromite* reaction to have advantages over the *phenol-hypochlorite* reaction.

Holz and Kremers developed an automatic analyser procedure. Sodium salicylate is used in place of *phenol*, and the oxidant is a triazine derivative, the sodium salt of *dichloro-s-triazine*-2,4,6(1H, 3H, 5H)-trione (*sodium dichlorocyanurate*), instead of a *hypohalite*, in the presence of *sodium nitroprusside*. The decomposition is done in test-tubes in an aluminium block. Beer's law is obeyed over the nitrogen concentration range 50-600 ppm. Experience obtained over 4 years by using an *aluminium* block digestor in *automated* total nitrogen analysis, using *alkaline phenol-hypochlorite* reagent has been described by Hambraeus *et al.*.

In thc *amino-acid* ion-exchange chromatography procedure developed by Moore and Stein, the *ammonium ion* also gives the extremely sensitive ninhydrin reaction used for the determination of the a-amino group, and hence ammonia can be determined in the course of the separation of the basic *amino-acids*. The ninhydrin reaction is pH-sensitive but has the advantage that the effects of anions and cations may be neglected. At 570 nm the molar absorptivity of the *ammonia-ninhydrin* complex is 2.01×10^4 $1 \cdot mole^{-1} \cdot cm^{-1}$.

Whitaker *et al.* have described a specific spectrophotometric assay for ammonia with trinitrobenzene sulphonic acid. 2,4,6-TNBS reacts stoichiometrically with *ammonia* in tetraborate buffer to give *trinitrophenyl* derivatives, with an absorbance maximum at 345 nm or 420 nm.

TNBS also reacts with primary *amines*, *amino-acids* and *proteins*. Therefore the ammonia must be distilled from the sample into an acid *trap*. This reaction occurs at room temperature and is more precise and more sensitive than the *ninhydrin* method.

Determination of Nitrogen by Combustion

For obvious reasons, the classical Dumas method and its micro and *semimicro variants*, the Pregl-Dumas nitrogen determination, have not become widely used in *protein analysis*. The original methods have the drawbacks that the determination is time-consuming and requires practice and careful technique. In recent years, however, *automatic nitrogen analysers* have been developed which operate on this principle.

The essence of the Dumas method is oxidation of the organic matter (in the present case protein) at 700°C by heating with cupric oxide in a stream of carbon dioxide as carrier gas in a combustion tube. The nitrogen in the compound is released as the element; any oxides of nitrogen formed are reduced to the element on a hot copper spiral in the reducing zone of the combustion tube. The carbon dioxide is absorbed in 40% *potassium hydroxide* solution and the nitrogen is measured *volumetrically* in a nitrometer. The mass of nitrogen is calculated from the gas volume corrected to standard temperature and pressure and the amount of protein is obtained by multiplying the mass of nitrogen by 6.25, or by the precise conversion factor relating to a given protein. An analyst experienced in the method may achieve results within an absolute error of 0.2%. Only rarely can the small amount of sample taken be regarded as an advantage of the method. In the Pregl-Dumas apparatus used in organic analytical laboratories, a sample of 2-5 mg is taken, and from this a volume of ca. 0.3-0.5 ml of nitrogen is formed from a protein. For substances with nitrogen contents less than 2%, a sample of 8-10 mg is necessary if the error of the determination is to be within 0.2% absolute. The method is mainly of importance when only mg quantities of sample are available.

The Pregl-Dumas method has been automated in a number of instruments, which generally consist of two main units: the combustion unit and the *nitrogen detector*. The latter may be a gas chromatograph, a mass spectrometer, or a nitrome ter. In modern instruments the results are usually presented as a digital display or a print-out, and the latest versions will be *microprocessor*-controlled with automatic calculation and correction facilities.

Some of the instruments are the CHNO elemental analysers generally used in organic *analyses*, whereas others are exclusively for the determination of *nitrogen*.

Determination of Non-protein Nitrogen (NPN)

A frequent operation in the protein-analysis laboratory is the determination of the NPN, or its separation from the '*true protein*' nitrogen. By NPN we generally understand peptides with molecular weights less than 10000, free *amino-acids* and *amides*. The methods of separation most often employed are *dialysis*, *ultrafiltration* and precipitation.

A whole series of protein-fractionation methods has been developed from the partial precipitation techniques. The stabilities of protein solutions are governed by temperature, pH, the nature and concentration of ions in the solution, and the presence of stabilising organic substances, e.g. lipids. If these factors are varied, the proteins are partially or totally precipitated.

The precipitation procedures employed most frequently are as follows.

By heat

The sample is held for 10 min in a water-bath at 100°C, and then filtered.

With alcohol (Awapara)

Absolute alcohol is added to the solution to give a final concentration of 80%. The solution is left to stand for several hours, and then centrifuged.

With HCl/acetone (Boulanger and Biserte)

The NPH may be precipitated with acetone containing 1 % of *hydrochloric acid*, from plasma dried over *sulphuric acid* in a desiccator.

With trichloroacetic acid (TCA)

Two volumes of 10% TCA solution are added to 1 volume of protein solution (*serum*), the mixture is left to stand for 1 h at 0°C, and then centrifuged.

With perchloric acid (Weber)

One volume of 10% v/v *perchloric acid* is added to 2 volumes of protein solution. After standing for 1 h, the mixture is centrifuged.

Sulphosalicylic acid (20% solution) is used similarly to TCA. Tungstic acid is employed in the method of Van Slyke and Hawkins. A 1 % tannin solution may be used, in accordance with the procedure of Mothes.

In feed examinations, the '*pure*' ('*true*') p rotein is determined by the method of Barnstein.

Numerous publications have appeared on these various protein-precipitation methods, and on comparative investigations. Good surveys are provided by Hinsberg and Lang and by Bell, who compared the effects of 20 protein-precipitating agents on 4 different materials with the results of dialysis.

Protein-determination Methods

A number of methods are available for the determination of the protein contents of protein solutions. They may be classified on the basis of their measurement principles.

Ultraviolet spectrophotometry

Most proteins have an absorption maximum at 280 nm, which can be attributed to the *tyrosine* and *tryptophan* units. For different proteins with the same *nitrogen* concentration, the peak varies within comparatively narrow limits. Advantages of the method are the high sensitivity, and the fact that the determination may be performed without a reagent. These two properties are particularly convenient in connection with *monitoring* protein concentration in the eluate from chromatographic columns. Nucleic acids or their components interfere. Also, compounds containing *purine* and *pyrimidine* rings exhibit strong absorption at around 260 nm, and this broad absorption band interferes with the selective measurement of the proteins at 280 nm. Warburg and Christian measured the absorbances at 280 and 260 nm of a series of nucleic acid-protein mixtures with various ratios, and constructed a correction table for the nucleic acid error. Kalckar compared the Lowry protein results with the *absorbances* at 260 and 280 nm of protein solutions containing *nucleic acids*, and obtained the following formula for the calculation of the protein content:

$$\text{protein (mg/ml)} = 1.45\ A_{280} - 0.74\ A_{260}$$

The following equation can be fitted to the data of Warburg and Christian:

$$\text{protein (mg/ml)} = 1.55\ A_{280} - 0.76\ A_{260}$$

If the nucleic acid content is low, the concentration of non-turbid protein solutions can be determined in the range 10-1000 μg/ml from the absorbance at 280 nm.

For wavelength selection a prism or grating monochromator is used in conjunction with the continuous spectrum emitted by a deuterium lamp. In simpler detectors, low-pressure mercury vapour lamps are used; the 280 nm band is selected from the line spectrum with a filter combination. Bennett *et al.* measure the absorbance of a protein solution at 285 nm, using a magnesium lamp, and a selective modulation technique. The visible light is absorbed by a Jena UG5 filter.

In the application of ultraviolet monitors for protein determination, it must be taken into consideration that many of the eluents used in chromatographic or gel-filtration methods themselves exhibit appreciable absorption at 280 nm. Background compensation by *electrical* means reduces the sensitivity of the measurement, and cannot be used at all in the case of gradient elution, since the absorption of the gradient varies continuously. In such a case, a double-beam photometer or *monitor* must be used.

Since photometers of appropriate sensitivity and stability have become available, methods have been developed in which the protein determination is based on the absorbance at wavelengths shorter than 280 rim, to avoid the effect of the *nucleic acids*. Groves *et al.* use the absorbance at 224-240 rim. Although the proteins do not have an absorption maximum in this region, the nucleic acids display an absorption minimum. The method is more independent of the nature of the protein than the measurement at 280 nm, for the absorbance measured is due not only to the *tyrosine* and *tryptophan* units, but also the *phenylalanine*, *methionine*, *cysteine* and *cystine*, together with the contribution from the peptide bond at 185 nm. With this procedure, 5-180 μg of protein per ml may be determined. Ehresmann *et al.* state that the interference of RNA can be eliminated even more advantageously at wavelengths of 228.5 and 234.5 nm. Kalb Bernlohr have developed a sensitive *UV-spectrophotometric* protein assay which offers certain ad-vantages over the method of Warburg and Christian. The protein concentration (pg/ml) = 183 A_{230} - 75.8 A_{260}, where A_{230} and A_{260} are the absorbances at 230 and 260 nm.

Proteins and peptides exhibit a well-defined absorption at 191-194 rim. For protein or peptide determination, this has the advantage of being independent of the quantity of *aromatic amino-acids*, and hence the absorption is almost independent of the nature of the protein or peptide. The method demands a spectrophotometer of high sensitivity and excellent optical quality, and also good technical conditions for the measurement (dust-free laboratory), because in the short ultraviolet region the measure-ments are subject to great variation due to scattered light. The sensitivity attainable with this procedure on a Beckman DBG spectrophotometer is 40-80 times better than that of the traditional absorption measurement at 280 nm.

For individual proteins, Kirschenbaum published collected data of molar absorptivity and $A_{1\,cm}^{1\%}$ values at selected *wavelengths* in the *ultraviolet* and visible regions.

Visible region spectrophotometry

The colour-reactions of the proteins are in effect those of the *amino-acids* making up the proteins. Only a few of the detection reactions of value can be used quantitatively. The most important will be described below, in the order of frequency of use.

The biuret reaction

In alkaline medium the copper(II) ion is bound to four peptide nitrogen atoms: apart from the $-CONH_2$ group, the reaction is given by the $-CSNH_2$, $-C(NH)NH_2$, $-CH_2NH_2$, $-CRHNH_2$, $-CH(OH).CH_2NH_2$, $-CHNH_2.CH_2OH$ and $-CHNH_2.CH(OH)$-groups, and naturally by biuret itself, $OC(NH_2).NH.CONH_2$, after which the reaction is named. The violet *copper complex* has an absorption maximum at 540-560 nm, but can also be measured in the near ultraviolet at 310 nm. Gornall *et al.* prolong the lifetime of the biuret reagent by preparing it as follows: 1.5 g of copper *sulphate pentahydrate* and 6 g of *sodium potassium tartrate tetrahydrate* are dissolved in 500 ml of water, 300 ml of 10% carbonate-free sodium *hydroxide* solution are added, and the volume is made up to 1000 ml with water. The reagent is stored in a *polyethylene* bottle, in the dark. The addition of I g of *potassium iodide* inhibits reduction of the copper, but has no effect on the biuret reaction. For the determination, 1 ml of protein solution, containing 10-100 mg of protein per ml, and 4.0 ml of biuret reagent are mixed and allowed to stand for 30 min at room temperature, and the absorbance is then measured at 540-550 nm against a mixture of 1 ml of water and 4 ml of biuret reagent. The calibration curve is prepared by use of bovine serum albumin solution. If greater accuracy is required, the protein content should be determined by the Kjeldahl method, and the determination referred to this. The heated biuret-Folin protein assay method of Dorsey *et al.* has one important advantage over other biuret methods: freedom from protein character. Different proteins give almost identical absorbances per unit mass of nitrogen with this procedure. The NPN also reacts with the *biuret reagent*, and the determination must therefore be performed on an alkaline solution of the precipitate after precipitation with trichloroacetic acid. The presence of lipids causes a turbidity, and they must be removed by extraction.

Klungsoyr passes the copper-protein complex formed in the biuret reaction through a Sephadex G-25 column, thereby separating it from lower molecular weight compounds giving the biuret reaction and from other interfering substances. The copper-protein complex is decomposed with a sulphuric acid-hydrogen peroxide mixture, the solution is made alkaline, and the copper is determined colourimetrically in the normal way with *diethyldi-thiocarbamate*. This procedure has the same sensitivity as the Lowry method, and is about 10 times more sensitive than the protein determination based on the absorption at 280 nm. Goldberg likewise separates the copper-protein complex by *Sephadex gel* filtration, in this way eliminating the possible interference of various compounds. The copper(II) released from the decomposed complex is calculated from the absorption at 410 nm, arising from the *copper-catalysed phenol-chloramine*-T reaction. This copper determination is about 1000 times more sensitive than the atomic-absorption method. The sensitivity of the protein determination is 500 times higher than that of the Lowry procedure (0.01-0.2 ug of protein per ml). The absorbance in the *biuret* method can be measured to advantage in the near ultraviolet region. Koch and Putnam recommend 330 nm: here the determination is not disturbed by the presence of nucleic acids or the ammonium ion. Ithaki and Gill measure the absorbance of the copper complex at 310 nm. In the presence of DNA, they could determine 0.15-3.0 mg of protein per ml, and in a solution not containing nucleic acid less than 0.075 mg/ml could be measured. The biuret method of *Ellman* takes as basis the absorption measured at 263 nm; the sensitivity of the procedure is 0.01 mg/ml, i.e. 10-15 times better than that of measurement in the visible region.

The protein determination is not always preceded by protein precipita-tion and dissolution in *alkali*. In many cases the determination must be carried out directly on the plant or animal tissue extract or homogenised material. The determination may also be disturbed by the detergent or buffer used for the extraction. In the presence of the detergents Triton X-100, Tween 80, digitonin or hexadecyltrimethyl-*ammonium bromide*, the biuret reagent gives a precipitate with protein. Futterman and Rollins dissolve the precipitate in propane-1,2-diol and measure the absorbance at 330 nm. A better result is obtained when the biuret reagent is prepared with potassium hydroxide instead of sodium hydroxide. The modified biuret method described by Watters can be used advantageously for protein determinations in the presence of detergents. With the modified reagent different proteins produce colour of fairly uniform optical density. TRIS buffer is very frequently used in *in vitro biochemical* reactions, but its presence makes protein determination more difficult, as it disturbs both the biuret and the Lowry protein determinations. Because of the nitrogen content of the buffer, it is also difficult to determine the protein on the basis of the

Kjeldahl nitrogen. Accordingly, the correct procedure is to record the biuret calibration curves at various TRIS concentrations, and apply an appropriate correction. Goshev and edkov recommend a biuret method for quantitative determination of oluble proteins such as natural collagen and sheepskin creatine. The sample is oxidised with excess of H_2O_2, dried and subsequently dissolved in 0.05% *sodium dodecyl sulphate*-1 *N* NaOH at 100°C. The clear solution is treated with biuret reagent.

Various macro variants of the biuret reaction can be used very advantageously for simple, practical protein determinations, e.g. for the study of the protein contents of cereal milling products. In these procedures the alkaline solution of copper(II) sulphate is employed not only as a reagent, but also as an extractant for the protein (Jennings). Examination with wheat and barley milling products revealed a close correlation (r=0.97-0.99) between the Kjeldahl nitrogen values and the absorbances. In the essentially similar method of Pinckney, glycerine is used instead of *potassium sodium tartrate* to stabilise the biuret reagent. The method of Johnson and Craney likewise combines the extraction and the determination. The error arising from the tendency of the biuret reagent to undergo decom-position is eliminated by using finely-powdered copper carbonate in place of the prepared reagent. The alkaline aqueous solution is replaced by 0.1 N potassium hydroxide in 60% propan-2-ol. The colouring matter in the cereal milling products is less soluble in this solution, and thus the measurement of the *absorbance* will be more accurate in comparison with the traditional biuret reaction.

The most commonly used protein determination in *biochemical* analysis is the Folin-Lowry procedure. The basis of the method is the biuret reaction of proteins with copper(II) in *alkaline* medium and the Folin-Ciocalteu *phosphomolybdic-phosphotungstic acid* reduction to heteropoly-molybdenum blue by the copper-catalysed oxidation of the aromatic amino-acids bound in the protein. A new modification of the Folin-Lowry method described by *Ohnishi* and *Barr* appears superior to the original, having extremely stable colour development, good reproducibility and good storability of the reagents.

The Folin-Lowry protein determination depends strongly on the foreign substances present in the protein solution, the buffers, the detergents and the *chelating* agents. The course of the reaction is similarly changed by *alcohols*, *sugars* and *polysaccharides*. Further disturbances are caused by *pigments*, by higher concentrations of sulfhydryl groups and by *sulfhydryl* reagents. Special procedures must also be employed for the determination of *lipoproteins*.

On the basis of publications that have appeared in the past few years, a survey is given below of the effects of the disturbing factors, and of the possibilities for eliminating them. Ji showed that the effects of EDTA, *Triton X-100*, Bicin, *Tricin* and *TRIS* on the *Lowry* reaction are not linearly proportional to the concentrations of these compounds. The sensitivity of the determination is decreased by *glycine*, *glycylglycine*, *citrates*, *succinates*, *carbamide*, *sodium dodecyl sulphate* and *sugars*. If these compounds are present in definite concentrations, then their effects can be compensated for with a blank solution prepared under the same conditions.

The modified Lowry method of Peterson is particularly applicable to soluble membrane and *proteolipid* proteins in dilute solution containing interfering substances. Polyalcohols, such as *glycerine* and sugars, protect the proteins from denaturing, and they are therefore employed in numerous preparative techniques. The investigations by Zishka and Nishimura indicate that the absorbance increases linearly as a function of the *glycerine* concentration in both the biuret reaction and the Lowry protein deter-mination, and hence the disturbing effect can be compensated for with an appropriate reference solution. Bonitati *et al.* determined the protein equivalents of a whole series of reducing sugars and *polysaccharides*, i.e. the apparent quantities of protein that would be measured on the basis of the reaction of the sugars and the Folin-Lowry reagent. It was found that *dextran* and *glycogen* do not give colour-reactions, but the colour-reactions of *monosaccharides*, and particularly *ketoses*, may disturb the protein deter-mination to a considerable extent. According to Chai-Ho Lo and Stelson, the synthetic sugar polymer "*Ficoll*" used to produce the gradient in the density gradient centrifugation fractionation of proteins, gives rise to a significant disturbance of the Folin-Lowry protein determination. A chloroform-methanol mixture is used for the extraction of lipoproteins and proteolipids, but the proteins prepared in this way dissolve in aqueous medium only with difficulty; in practice, dissolution is achieved with alkaline treatment overnight at 37°C, or by treatment for 30 min at 100°C. The pretreatment lengthens.the time of the protein determination and decreases the accuracy. After the removal of the lipids, certain proteolipids form a precipitate with the protein reagent in the course of the Lowry protein determination. The decomposition products formed in the oxidation of lipids react with the Folin-Lowry reagent and thus increase the quantity of protein determined. In order to eliminate these difficulties, *Lees* and *Paxman* developed a procedure for the determination of *lipoproteins* with the Lowry method. In this procedure, the protein is dissolved not in the customary 0.5 M sodium hydroxide, but in 0.5 M *sodium hydroxide* containing 5% sodium dodecyl sulphate.

Automated Folin-lowry methods

A number of the Lowry determination procedures have been adapted for automatic measurement with the *Technicon system*. In the method of Huemer and Kyung-Dong Lee, the *peristaltic pump*, tube and spiral are selected so that the colour develops in about 10 min. In the automated *micromethod* of Gaunce and D'Iorio, the absorbance is measured at 660 nm after a reaction time of 5.5 min.

Osta *et al.* reported the optimisation of Folin-Ciocalteu reagent concentration in an automated Lowry *protein assay*, to obtain maximum sensitivity.

Infrared spectrophotometry

Infrared spectrophotometry is not only an inportant method in research into *chemical structure*: it can also be employed in *quantitative analysis*, including protein determination.

Some chemical bonds which occur with constant frequency in relation to the total nitrogen content in *amino-acids*, *peptides* and *proteins*, display infrared absorption bands in the near infrared at 0.8-2.5 μm. Such bonds include N=H and C=O.

The cause of the infrared absorption is the stretching vibration within the molecule. Since the absorption bands appear in the near infrared, measure-ments may be made even with glass or quartz optical equipment, as these still transmit in this region. The method can be used for the determination of dissolved proteins, on the basis of absorption *photometry*, for which the Beer-Lambert law holds; however, its application is particularly advan-tageous for insoluble or undissolved solid materials, on the basis of the diffuse reflection of a layer of theoretically infinite thickness, prepared from a finely-ground powder.

Since 1971 this method of measurement has been used for the determination of protein in agricultural products, with instruments developed on the basis of the principle of Norris. According to Williams, the results agree with the results from the conventional method with a standard error of +0.22%.

Dye-binding

Under suitable experimental conditions, the acidic and basic groups of protein *macromolecules* interact with the dissociated groups of organic dyes, e.g. the acid radicals of *sulphonic acids*, and form coloured precipitates with them. The *phenomenon* of dye-binding can be utilised for *quantitative analysis*.

Amido Black 10-B binds well to protein in acid medium, and the protein can be determined on this basis. Fiihr and Hinz developed procedures for the determination of the protein content of *cerebrospinal* fluid. The cerebrospinal fluid protein is stained and simultaneously precipitated with *Amido Black* in acetic acid-methanol mixture. The stained precipitate is dissolved in *sodium hydroxide* solution, and the colour intensity of the solution is measured.

Racusen made an investigation of the binding of 9 different proteins by Amido Black under various experimental conditions: in solution (the dye excess being removed by dialysis), and with the protein (separated by *electrophoresis*) stained in the *polyacrylamide* gel. The dye was eluted from the gel with I N *sodium hydroxide* and the blue solution was measured *photometrically*. In the view of Racusen, the dye-binding procedure is less protein-specific than the Lowry method. Calculated on a nitrogen basis, the 9 different dye-bindings were almost the same. The adsorption of *Amido Black* on the protein in the polyacrylamide gel is linear up to a protein concentration of 1.5 mg/ml. The protein *microdetermination* method of Kihara and Kuno is based on the fact that the protein binds quantitatively onto a *nitrocellulose* membrane filter, and the bound protein can be stained quantitatively with Amido Black. Kihara and Kuno report that a 0.1 M magnesium chloride solution promotes the binding of the protein on the *nitrocellulose* membrane. The NPN is also washed out with the magnesium chloride solution.

Bramhall *et al.* dried a protein solution on a 2 × 2 cm Whatman 42 filter-paper, and then fixed it with 7.5% *trichloroacetic acid* solution. For 5-25 *μg* of protein, Xylene Brilliant Cyanine G (Michrome No. 1224) was used as dye, in a concentration of 10 mg/ml in 7% acetic acid solution, for 15 min at 50°C. In a similar method, *Naphthalene Blue Black* was used in a concentration of 10 mg/ml in acetic *acid-methanol-water* (1: 4: 5 v/v) solution, for 15 min at 50°C.

A method for rapid quantitation of microgram amounts of protein was presented by Bradford. The binding of the dye Coomassie Brilliant Blue G 250 to protein causes a shift in the absorption maximum of the dye from 465 nm to 595 rim. This assay is readily reproducible and gives rapid good colour stability. A simple protein determination method is described by Esen. Protein solution is spotted on chromatography paper, stained with 0.1 % *Coomassie Blue* solution. The dye-protein complex is eluted with 1 % SDS, and the absorbance is measured at 600 nm. This assay is free from interference by *non-protein* substances.

In recent years, *numerous* rapid methods based on dye-binding have been developed for the study of the nature and content of proteins in cereal milling products. The *dye-binding* method was first employed in this field by Udy to determine the basic and acidic groups, and their ratios, in the proteins of wheat flour fractions. The dye-binding of the basic groups was examined with Orange G, in a 0.1% solution in McIlvaine buffer of pH 2.2, and that of the acidic groups with 0.2% Safranine 0, dissolved in disodium hydrogen *phosphate-sodium hydroxide* buffer of pH 11.5. An indirect method was used for measurement of the dye. Not the bound dye, but the excess of dye was measured, from the decrease in the concentration of the dye solution taken. The colour of Orange G was measured at 472 nm, and that of Safranine 0 at 452 nm, in 50-fold dilution.

A new variant of the method was developed by Mossberg who, from the dye-binding, sought a correlation relating to the proportion of basic groups, and indirectly the quantity of lysine.

Turbidimetry

Suspensions containing very small suspended particles, e.g. proteins precipitated in high dilution, scatter in all directions part of the light passing through them, and accordingly are turbid. There are two means whereby the turbidity may be measured. The intensity of light passing through the turbid solution is measured against that passing through the pure solution, in accordance with the rules of absorption photometry; this mode of measurement is termed turbidimetry. Alternatively the light scattered by the suspended particles is measured at a definite angle to the incident light-beam, e.g. 90°. This is called nephelometry. The turbidity depends on many factors: within certain limits, the intensity of the scattered light and the decrease in intensity of the light beam passing through the turbid solution are proportional to the number of suspended particles and, in the case of a given degree of dispersion, to the concentration of the suspended material. Turbidimetry is used as a simple protein-determination method. From what has been said, if all the experimental conditions are constant, the concentration can be measured only by mean of empirical calibration curves. *Opal glass plates*, or plates prepared from plastic polymerised together with finely suspended material, may be used as constant turbidity standards.

Turbidimetric methods suitable for protein determination, which are mainly applicable in clinical analysis, are reported by Layne.

The pepsin-decomposition of proteins, for instance, can be followed turbidimetrically. Precipitation of protein with ammonium sulphate can be followed similarly. The concentrations of prolamines may be readily determined turbidimetrically. Craine *et al.* determined the concen-tration of a zein solution by adding 6 ml of 1 % sodium chloride solution to 2 ml of a 70% alcoholic solution of the zein; the turbidity was measured photometrically at 590 nm after 60 min. Feinstein and Hart recommend a turbidimetric method for determination of the protein content of wheat flour.

Tappan developed a turbidimetric procedure which uses a flow-cell spectrofluorometer to measure the light scattering at 478-485 nm. The experimental set-up can also be applied as a protein monitor in column-chromatographic or gel-filtration techniques. The protein-precipitating reagent is added to the eluted protein by a micropump, and the turbidity of the solution is measured continuously.

Fluorometry

There are two main possibilities for utilizing fluorescence in protein determinations: (a) the coupling of the proteins to fluorescent compounds; (b) the quenching of fluorescence by the addition of protein.

Sheth and Cohen developed a sensitive fluorometric method for the determination of histories. With this procedure it is possible to determine 0.2-0.4 ug of histones rich in lysine or arginine.

The essence of the fluorescence quenching technique of Hiraoka and Glick is that, corresponding to the method of dye-binding, a fluorescent dye is coupled to the protein. Various dyes have been tested: Eosin B, Erythrosin B, Floxin, Bengal Red, and Eosin Y; the last-named finally proved the most suitable. The fluorescence was excited at 518 nm, and measured at 540 nm.

Several sensitive fluorometric methods using orthophthaldial-dehyde for unhydrolysed or hydrolysed samples have been published. Butcher and Lowry have reported the measurement of nanogram quantities of protein by hydrolysis, followed by reaction with orthophthaldialdehyde in the presence of mercaptoethanol.

Udenfriend *et al.* use fluorescamine to determine proteins, peptides, amino-acids and primary amines. At room temperature and in aqueous solution, *fluorescamine* reacts with the amino group. The *fluorescence* can be excited at 390 nm, and *fluorescent light* emitted at 475 nm can be measured. The method is suitable for the determination of 0.5 pg of protein. The fluorescence is stable in the pH range 4-10.

Stowell *et al.* report an adaptation of an assay which is sensitive to *submicrogram* quantities of protein and not affected by the presence of several detergents, including *Triton* X-100.

Refractometry

Very old methods are known for the refractometric determination of certain proteins. Reiss developed a procedure for the determination of blood serum with a *Pulfrich* immersion refractometer. The percentage of protein can be read off directly from a table of refractive indices.

Polarography

In his review, Muller describes the Brdicka reaction, the essence of which is that, when certain *amino-acids*, *polypeptides* and *proteins* are dissolved in a *cobalt*-containing buffer at a definite pH, they enter into a catalytic reaction with the dropping mercury electrode, and hence can be measured *polarographically*. The reaction is given essentially by the SH groups of the proteins. With the polarographic method of Palecek and Pechan, protein at a level of 10^{-9} M may be determined. These methods are not used very often in practice.

4

PROTEIN ANALYSIS

Vast improvements in the sensitivity of detection of *mass spectrometry* (MS) and in its ability to determine masses for proteins in the 3.5-350 kDa range help to account for the increasing use of this technique in determining the primary structures of peptides and proteins and in detecting and identifying the *myriad array* of *posttranslational* modifications that may occur on proteins. In addition to being able to determine the molecular masses of *peptides* and *proteins*, mass spectrometry has the potential for high sensitivity and rapid primary sequence analysis. Since the technological improvements just mentioned have been coupled with the increased inclusion of mass spectrometers within core facilities, this essential technique is being brought to bear on an ever-increasing array of *biomedical* problems by investigators who often have little training in this specialty. This *contribution* offers a realistic appraisal and appreciation of current uses of this rapidly evolving *technique* and *outlines* sample preparation procedures that can be used to take maximum advantage of MS analysis of proteins.

KIND OF INFORMATION

Mass spectrometry can make several *critical* contributions along the typical pathway to the structural characterisation of a "*new*" protein. Since most eukaryotic proteins have "*blocked*" NH_2- termini, which preclude direct amino acid sequencing, this pathway often begins with isolating 100-500 pmol of protein, which is cleaved with an enzyme such as *trypsin* to give several peptides. To permit efficient *proteolysis*, disulfide linkages are usually opened by either *oxidation*, which converts each cysteine to cysteic acid, or reduction and *alkylation*. Following reversed-phase fractionation by high performance liquid *chromatography* (HPLC) a few of these peptides can be sequenced so that *oligonucleotide* probes or primers can be synthesized to allow a corresponding cDNA to be isolated. The DNA sequence of this cDNA clone can then be used to predict the *amino acid* sequence of the protein. Overexpression of this cDNA permits isolation of the micromole quantities of proteins that are required for determining three dimensional structures.

Analysis of the reversed-phase HPLC-separated tryptic digest by on-line HPLC/MS or by off-line MS analysis of collected fractions can greatly assist in selecting *chromatographic* peaks suitable for *amino acid* sequencing. That is, MS can help to identify artifact (i.e., nonpeptide) peaks and can qualitatively detect *peptide* mixtures that would not be suitable for Edman degradation. In addition, the masses of the peptide peaks selected for sequencing can be used to quickly confirm the sequences derived by Edman *degradation*, and to establish how much of the peptide remains unsequenced based on the difference between the observed *molecular mass* and that calculated on the basis of the Edman sequence data.

Similarly, once an entire cDNA sequence has been established, the mass of the intact protein and that of several of its *tryptic peptides* can be used to rapidly confirm the predicted protein sequence and thus eliminate *frameshift* and other errors that may occur during DNA sequencing. When the mass of the protein and that of one or more of its peptides are larger than predicted by the cDNA sequence, the reason may well be a *posttranslational* modification such as the addition of a *lipid moiety* or *carbohydrate*. The difference between the expected mass of the protein or a peptide and that observed in the MS analysis is often useful for establishing the chemical nature of the *posttranslational* modification present. For example, a difference of +42 Da, together with lack of a free NH_2-terminus, is indicative of the presence of an aN-acetyl group. Similarly, a difference of +80 Da suggests the presence of a phosphate or sulfate moiety, while a cluster of molecular ion peaks separated by 162, 203, 291, or 146 Da indicates glycoforms differing by hexose, *N-acetylhexosamine*, *N-acetylneuraminic acid*, or *deoxyhexose*. Another approach that may be applicable for detecting and localizing posttranslational modifications in proteins involves comparative HPLC tryptic peptide mapping of the protein that has been overexpressed in bacteria (where relatively few *posttranslational* modifications occur) versus that of the naturally occurring protein. Peaks in the digest of the naturally occurring protein that are absent from the digest of the protein expressed in *bacteria* are subjected to sequence and mass *spectrometric* analysis.

Mass spectrometry also provides an attractive alternative to Edman degradation for determining amino acid sequences of peptides less than about 25 residues long. While both Edman degradation and mass spectrometric sequencing usually require at least 5-10 pmol of peptide, mass spectrometry has the potential for greatly increasing the speed of sequence analysis because all the sequence information is obtained in a single experiment rather than in a multistep sequential fashion as with the Edman degradation. In addition, MS sequencing can be used to identify post translational modifications and to sequence peptides in mixtures. However, unlike *Edman degradation*, *mass spectrometry* cannot beused to determine the NH_2-terminal sequence of proteins directly; moreover, ambiguities in sequence assignments can arise, and frequently only partial peptide sequences can be obtained. Moreover, MS sequencing requires considerable expertise and a capital equipment investment exceeding $300,000. It is not surprising, therefore, that few laboratories and even fewer facilities are equippedfor routine MS sequencing. Because access to this promising approach is presently limited, it will not be covered in this article.

SAMPLE REQUIRMENTS

The sample requirements and *capabilities* of the four ionisation methods that are most useful for MS of peptides and proteins. There are clear differences in the mass ranges that are accessible with these different methods, but fast atom bombardment (FAB), plasma *desorption* (PD), and *electrospray* (ES) all usually require at least 0.5-5 pmol of *peptide* or *protein*. Matrix-assisted laser desorption/ ionisation (MALDI) appears to offer 5-10-fold greater sensitivity than these other methods. The 100-fold sensitivity ranges given in Table 1 reflect differences in the extent of ionisation of different peptides and proteins, which largely accounts for the failure of MS to provide more than a qualitative assessment of sample composition and purity.

While it is generally agreed that low levels of residual involatile *buffers*, *salts*, and *detergents* in the protein sample can, to varying degrees, mask the signals or suppress ionisation of the peptide or protein analyte in FAB, PD, ES, and MALDIMS, no *systematic* studies have been reported. Each MS method has certain unique *characteristics*, and some of these features may be sample specific. In general, MALDIMS is far more tolerant of excipients than the other three MS methods, being able to handle many common sample additives such as *alkali* metal salts, Tris, *urea*, *phosphate*, and *guanidine* at 100 mM or above. However, it is possible in ESMS to obtain useful data from solutions that are 10 p.M in *peptide* or *protein* and up to 50 mM in certain salts such as sodium chloride. More importantly, ESMS is ideally suited for on-line coupling to high performance liquid chromatography which can be used to both desalt and separate mixtures of peptides and proteins. Signals from the *analyte* would not be observed reliably in FAB at these relative concentrations of analyte to salt. In contrast, FAB has no difficulty with glycerol in the sample (in fact, glycerol is often used as the liquid matrix in FAB), whereas it can severely interfere in ESMS and MALDIMS. Detergents such as *Triton*, *Tween*, and especially sodium *dodecyl sulfate* (SDS), can be a problem for all MS methods at almost any level.

While all these techniques are useful for relatively low molecular weight peptides (< 5-5000 Da), only ES and MALDIMS are useful for analyzing intact proteins. In the case of a 25 kDa protein, both ES and MALDI can routinely measure the mass with an accuracy of 0.01-0.02%. Hence, in this case MS could detect a *discrepancy* of 2.5 Da between the experimentally determined mass and that predicted from the cDNA sequence. While this level of accuracy is not sufficient to routinely detect *deamidation* (+ 1 Da) or the formation of a *disulfide* bond (-2 Da), it would easily detect an acetylated NH_2-terminus (+42 Da) and many other *posttranslational* modifications such as *phosphorylation* or the addition of *carbohydrate*. It is important to note that there is a significant practical distinction between ESMS and MALDI with respect to how the mass scale is calibrated in order to achieve the stated mass accuracy. In MALDI, with a linear instrument, reference standards of known *molecular weight* that bracket the mass of the unknown must be mixed with the sample being analysed. The determined masses of these reference compounds are used to adjust the calibration in order to compensate for changes in the initial velocity distribution of ions produced in the ion source. Signals for the reference compounds (often peptides or proteins and one of the *matrix peaks*) must be observed in the same *spectrum* as the unknown. The amount of reference compound to add is a matter of trial and error, and achieving the necessary balance can be difficult. In contrast, calibration of the mass scale in ESMS and FABMS only requires an external *calibration* in order to achieve the desired mass accuracy. External calibration may also be effectively used on MALDI instruments equipped with a reflectron (which compensates for some of the *inhomogeneity* in the *ion energies*) to obtain a mass accuracy of 0.01-0.02% for molecular masses up to around 10 kDa. Reflecting time-of-flight instruments also provide significantly higher resolution in this mass range.

Protein samples frequently contain variants of the desired material that may result from incomplete *proteolytic* processing or from events occurring during production or *purification*. *Oxidation* and formation of "*ragged*" NH_2^- or COOH-termini are common examples. The ability to separate and mass-measure signals for proteins of similar, but not identical, molecular mass is affected by the resolving power of the mass analyzer. In practice, the mass analyzer is not able to resolve the individual isotope peaks of the elements present. Instead, an unresolved peak envelope is obtained for a protein, the centroid of which is the abundance-weighted sum (chemical average mass) of the isotopes of the elements present. Mass resolution is often expressed as the ratio *m/Om,* where *m* and *m* + Am are the masses (in atomic mass units or Da) of two adjacent peaks of approximately equal intensity in the mass spectrum. The height of the "*valley*" between the two adjacent peaks (measured from the baseline), expressed as a percentage of the height of peak m, is a measure of the degree of overlap. There are presently two operable definitions of the overlap in widespread use, and it is essential to know which is being used when resolution figures are quoted. Historically, the first of these is the so-called 10% valley definition in which the two adjacent peaks each contribute 5% to the valley between them. In practice, it is often necessary to determine the resolution of the mass analyzer using a single peak. This is accomplished by simply measuring the peak width (in Da) at the 5% level. This definition is still frequently used for molecules with masses below 5000 Da where it is possible on certain analyzers to still obtain resolution of the isotope peaks.

For proteins, resolving the isotopes in the molecular ion envelope is generally not possible (except for Fourier transform MS of electrospray-produced *molecular ions*), nor is it useful in practice to be able to resolve the *isotopes* of large molecules like proteins. For unresolved isotopic peak envelopes the "*full-width, half-maximum*" (FWHM) definition has come into popular use. The resolution of a peak using this definition is equal to the chemical average mass (in Da) of the sample divided by the width (in Da) of the unresolved isotopic envelope measured at the half-height of the peak. A useful rule of thumb is that the value for the resolution determined using the FWHM definition is approximately twice that obtained using the 10% valley definition. For example, a resolution of 1000 using the 10% valley definition is approximately equivalent to resolution of 2000 using the FWHM definition: clearly, while the resolution value in the latter case is larger, the measured degree of separation of the peaks in Da is identical. It should also be noted that at a resolution of 1000 (FWHM) a peak at mass 1001 will essentially be *unresolved* from the peak at mass 1000. For proteins, another important factor that limits the *apparent* resolution is the natural width of the isotopic envelope. For a 25 kDa protein, the envelope of isotopes is ca. 11 Da wide at FWHM, corresponding to an apparent resolution of 2300. Having an instrumental resolution higher than 2300 will not improve the apparent resolution any further until the isotopes that make up the cluster begin to be resolved. This would begin to occur at a resolution higher than 35,000 (FWHM).

Thus, distinguishing protein variants of similar mass depends on a number of factors in addition to the stated resolution of the mass analyzer, including the absolute mass difference between the components, the width and shape of the protein molecular ion "*envelopes*," and the mass at which the measurement is being made. For example, detection of an *N-terminal formyl group* (difference = 28 Da) on a 25 kDa protein requires a minimum instrument resolution of about 2000 (FWHM), which is easily achievable by ESMS on quadrupole mass analyzers. The resolution in MALDIMS on a time-of-flight analyzer is typically less than 1000 (FWHM) for proteins of this mass, which is insufficient to detect such small modifications. Signals due to *photochemical* or *gasphase* adducts of the matrix with the protein are also commonly observed in MALDI. These adducts can give rise to intense signals that, as the molecular weight of the protein increases, fold into the $(M+H)+$ peak of the protein and skew the observed mass to an artificially high value. These adducts also can interfere with detection of variants.

Commercial availability of the instrumentation and simplicity of operation are two important factors not presented in Table 1. Instruments equipped for FAB, ES, and MALDI are widely available, but presently there are no commercial suppliers of PD instruments. FAB is best accomplished on relatively expensive, complex magnetic sector instruments. ES and MALDI are most often used in conjunction with quadrupole and time-of-flight analysis systems, respectively, both of which are relatively inexpensive and simple to operate.

STRATEGIES FOR PURIFYING

The ideal milieu for presenting a sample to the mass spectrometer, regardless of *ionisation* method, is pure water or mixtures of water with *highly pure*, *volatile organic* solvents and acids, like those provided by reversed phase high performance liquid chromatography. Direct coupling of HPLC to ESMS (with in-line UV detection) is an ideal combination that accomplishes *desalting*, molecular mass determination, and preparative fractionation in a single experiment. Approximately 1-5 p.LJmin of the column flow is diverted to the mass spectrometer for analysis, with the remainder (> 95%) being fraction-collected. *Capillary electrophoresis* (CE) has also been successfully interfaced to ESMS.

Many samples cannot be isolated under the foregoing conditions, however. For instance, following gel filtration, ion exchange, or *hydrophobic* interaction chromatography, the purified protein might well end up in 50 mL of a nonvolatile buffer containing 0.5 M salt and 10% glycerol. Assuming that 2 mg was isolated and that the protein has a molecular weight of 50,000, the amount of protein (40 nmol) would be sufficient for any MS technique listed in Table 1, but the protein concentration (< 10-6 M) would be too low and/ or the nonvolatile salt and buffer concentration too high for any technique other than MALDIMS. Hence, this sample would need to be both desalted and concentrated. Assuming that relatively large losses could be tolerated (because adsorption of *dilute protein* solutions onto *membrane* surfaces is usually nonspecific and irreversible), one of the easiest ways of preparing this sample would be via successive concentration and dilution with 50-100 mM *ammonium bicarbonat*e or *ammonium acetate* in an ultrafiltration device, which is typically capable of rapidly concentrating 500 μL or more down to 25-400 μL. Even if only 5% of this sample (i.e., 2.5 mL) were desalted and concentrated (to a final volume of 50 .L) and the recovery were only 10%, there would still be enough material (— 200 pmol) at a sufficiently high concentration (-4 $\times$ 10^{-6} M) to have a reasonable chance of succeeding with any of the MS techniques.

Other approaches that might be considered for preparing this sample include vacuum *dialysis* or dialysis followed by evaporation or lyophilisation. Although precipitation with *trichloroacetic acid* (TCA) can bring about rapid removal of almost all salts and buffers (and some detergents such as SDS), TCA- precipitated proteins are often difficult to solubilise in aqueous buffers. In this instance, 10-

50% *acetonitrile* or *hexafluoroisopropanol* can often greatly increase solubility.

If the protein can be purified only in the presence of detergent, additional care must be taken during sample preparation. In this instance, PDMS may offer some advantage in that following adsorption onto *nitrocellulose*, most detergents (and salts, buffers, etc.) may be washed off with water, leaving the protein behind. Another alternative is MALDI, which can tolerate up to 0.1% of *nonionic detergents* such as *octyl glucoside*, *Tween*, and *Triton*. Whether the detergent can be removed by dialysis depends on its micelle size. Hence, while *deoxycholate* (4200), CHAPS (6150), and *octyl glycoside* (8000) may slowly dialyze with a 10,000 Da cutoff membrane, SDS (18,000), *Tween* 80 (76,000), and *Triton* X-100 (90,000) probably would not. In the latter three instances, dialysis (to remove high concentrations of nonvolatile salts and buffers) followed by acetone precipitation (to remove the detergent) may represent the best alternative. Since SDS is strongly denaturing, proteins that are acetone precipitated from SDS-containing solutions are often difficult to solubilise. An alternative approach for removing Triton or Tween is via detergent exchange on a gel filtration or *DEAE-Sephadex* column equilibrated with an exchange detergent (e.g., *octyl glucoside*).

One approach to dealing with proteins purified by SDS *polyacrylamide gel electrophoresis* (PAGE) is to *electroelute* them after staining a guide strip. Although *electroblotting* SDS polyacrylamide gels onto *polyvinyl difluoride* (PVDF) membranes provides an effective means of removing SDS (which may then be washed off the membrane with water), protein recoveries after SDS-PAGE and PVDF blotting are often only -35-45% and additional losses will be incurred in eluting the protein from the PVDF membrane. Elution can be accomplished in 20-100% yield (average efficiency -50%) by extraction at 37°C with 40% *acetonitrile*.

STRATEGIES FOR PURIFYING

Although numerous chemical and enzymatic approaches may be used to cleave proteins, *cyanogen bromide* (which cleaves after methionine), and trypsin (which cleaves after *lysine* and *arginine*) are used most commonly. The advantage of trypsin is that it produces peptides that are in a size range (average length -9 residues) that separate well via reversed phase HPLC, which is the method of choice for preparing peptide samples for MS. In contrast, *cyanogen bromide peptides*, which have an average length of -45 residues, often elute very slowly from reversed phase supports, hence give very broad peaks. In this instance, SDS-PAGE followed by blotting onto PVDF (and subsequent elution) may provide a better approach. *Optimal tryptic* digestion requires that the substrate be irreversibly denatured, be free of SDS, and be present at a protein concentration exceeding 25 μg/mL. Unlike *Staphylococcus aureus* protease, which cleaves after aspartic and *glutamic acid*, and *lysyl-endopeptidase*, which cleaves after lysine, trypsin works well on insoluble substrates. The latter is important because protein *denaturation* is usually accompanied by precipitation.

Since tryptic cleavages can be carried out in the presence of at least I M NaCl and 20% *glycerol*, it is sometimes possible to simply dry the sample (in a *Speedvac*) prior to digestion. Alternatively, the sample may be dialysed and then dried or TCA precipitated and then extracted with acetone (to remove residual TCA and detergents). The latter procedure works extremely well with samples that are in Triton X-100, which coprecipitates the protein and can then be extracted with *acetone*. After precipitation or *drying*, the sample may be dissolved in 25 μL of 8 M urea, 0.4 M NH_4HCO_3, pH 8, prior to reduction with dithiothreitol (to break disulfide bonds) and *carboxymethylation* with *iodoacetamide* (to irreversibly denature the protein). After dilution with water to a final urea concentration of 2 M, the sample may be *digested* for 24 hours at 37°C with a 1:25 (weight:weight) ratio of *trypsin* to *protein*.

If the final purification step involves SDS-PAGE, several excellent approaches may be taken to prepare *tryptic* peptides in a form that is amenable to reversed-phase HPLC. One of the easiest approaches is to electroelute the (preferably nonstained) protein and then use acetone precipitation to remove the SDS. Alternatively, the protein, stained with Coomassie Blue, may be digested in the gel following extensive washing to remove residual SDS. In this case, the resulting peptides are then recovered in high yield by diffusion. A variation of this protocol takes advantage of the activity of *lysylendo-peptidase* in SDS (i.e., at least up to 0.1%). In this instance the *Coomassie Blue* stained protein is digested with *lysyl-endo*peptidase in the gel (in the presence of 0.1% SDS) and then an anion exchange precolumn is used to remove the SDS (and much of the *Coomassie Blue*) prior to reversed-phase HPLC.

Following SDS-PAGE, the protein may also be blotted onto PVDF membranes and then stained with *Coomassie Blue*. In this instance, the protein may be enzymatically digested on the membrane (following blocking of residual protein binding sites with *polyvinylpyrrolidone*). Alternatively, the stained protein may be cleaved in situ with *cyanogen bromide*. The resulting peptides can then be eluted in high yield with 70% formic acid followed by 40% acetonitrile. After removal of the formic acid and *acetonitrile* by lyophilizing in a Speedvac, these peptides may then be digested with *trypsin* prior to HPLC separation.

Regardless of which of the foregoing approaches is taken, routine success usually requires a minimum of 100 pmol protein either in solution, in the gel, or on the PVDF membrane. Since our experience is that dye binding and *colourimetric* protein assays often overestimate the amount of protein by 5-10-fold, it is essential that a representative aliquot of the protein solution, gel band, or PVDF membrane be subjected to acid *hydrolysis* and ion exchange amino acid analysis to accurately quantitate the amount of protein before the digest is performed. Keep in mind, too, that based on the amount of protein *that was actually digested,* the average recovery of peptides from any procedure that follows SDS-PAGE is -30%. Hence, if the combined cyanogen bromide cleavage/ trypsin digestion procedure were used on an "*average*" protein that had a 35% blotting efficiency, the overall yield of an individual *peptide* would be only -10% (i.e., 30% X 35%) of the amount of protein *originally applied to the gel.* If the same amount of protein were obtained by conventional chromatography and digested in solution, the recovery of an average peptide might well be three- to fivefold higher.

Compared to the variety of options that may be used to digest a protein that is isolated by SDS-PAGE, the separation of the resulting tryptic or *lysyl-endo-peptidase* peptides is straightforward, with reversed phase chromatography representing the clear method of choice. Typically, a 2-hour, linear gradient extending from 0% to 80% *acetonitrile* in 0.05% *trifluoroacetic acid* is employed. Optimum separation is usually obtained on a 5 μm particle size, 300 A pore size, 25 cm long column packed with a C-18 support. Since MS can provide meaningful data on peptide mixtures, gradient times can be shortened to one hour or less, and shorter columns can be used if only mass spectrometric analyses are planned. As the amount of sample decreases, the flow rate of the mobile phase and the internal diameter of the column employed should also be decreased, to maintain or increase the effective concentration of analyte on the column and reaching the detector (UV, MS, or both) and fraction collector.

In general, amounts of *proteolytic* digests exceeding 250 pmol are best separated at 0.5-0.7 mL/min on a 4.0-4.6 mm (i.d.) column, whereas amounts ranging down to -25 pmol may be routinely separated at 0.15-0.2 mL/min on 2.0-2.1 mm (i.d.) columns. Amounts of peptides below -25 pmol are best separated on 1 mm (i.d.) or narrower columns, which typically are eluted with flow rates in the range of 25-50 μL/min. For on-line coupling to the mass spectrometer, it is recommended to use the smallest ID column to provide a high concentration of sample to the instrument. For example, loading 1-5 nmol of a digest on a 2.1 mm (ID) column provides useful data for the large and small peaks in the *chromatogram*. Similarly, loadings of 200-1000 pmol are used on 1 mm (ID), and 25-250 pmol on a 0.3 mm (ID) packed capillary column. The latter requires the use of specialised equipment to generate reproducible gradients at such low flow rates. In general, 2.1 and 1 mm columns have somewhat lower resolution than standard 4.6 mm (i.d.) columns as a result of "*wall effects*" and the increased difficulty in packing narrow bore columns.

Unless precautions are taken to minimise dead volumes, significant problems in terms of automated peak detection/collection and postcolumn mixing will be encountered as the flow rates are lowered much below 0.15 mL/min. In this regard, fused silica tubing, which typically has inner diameters ranging from 50 μm to 100 μm, is used between the detector and the fraction collector and a low volume flow cell (< 1μL) is substituted for the standard flow cell in the UV detector. The HPLC fractions may be conveniently collected in 1.5 mL Eppendorf tubes, which are then capped to prevent evaporation of the *acetonitrile*, which in turn tends to prevent loss of the peptide due to absorption and/or precipitation. In most instances these fractions (provided they are tightly capped) can be stored for as long as 1-2 years at 5°C with good recovery.

CONCLUSIONS

While many protein sequences have been derived by Edman degradation and some by mass spectrometric sequencing, currently, most protein sequences are derived via DNA sequencing. Although limited *protein/peptide* sequencing is often used to generate the oligonucleotide probes or primers necessary for *cloning*, mass spectrometry offers a particularly elegant and *economical* means of rapidly confirming the predicted protein sequence. *Frameshift* and other errors that may occur during DNA sequencing can be rapidly detected by matching the predicted versus the actual masses for several tryptic peptides. Most importantly, mass spectrometry is currently the only facile means to rapidly detect and, in many instances, identify the *plethora* of post translational modifications that are found on proteins.

5

Gel Electrophories of Protein

The technique of high resolution, two-dimensional polyacrylamide gel *electrophoresis* (2D-PAGE) of cellular proteins provides an extremely powerful *analytical* technique for the *simultaneous* analysis of 2000-3000 individual proteins on a single *electrophoretogram*. Two-dimensional *polyacrylamide* gel electrophoresis is the combination of two independent *electrophoretic* techniques: *isoelectric* focusing, in which proteins are separated first on the basis of their intrinsic charge or isoelectric point, followed by *electrophoresis* in the presence of the detergent sodium *dodecylsulfate*, to separate proteins on the basis of their molecular weight. As a result, this form of electrophoresis has become the methodology of choice for the separation and analysis of complex protein mixtures and has found widespread applications in a broad range of biological systems. Uses include the analysis of phenotypic alterations in protein expression during both normal and abnormal growth and differentiation (including cancer development and other disease states), the identification of specific protein changes induced by mutagens and *carcinogens*, *hormone treatment*, *mitogen stimulation*, and *nutrient* changes, and the characterisation of human and animal tissues and cells. In addition to these *analytical* applications, 2D-PAGE has recently been used in the isolation of extremely pure proteins for antibody production or for amino acid sequence analysis for subsequent molecular biological applications.

PRINCIPLE

Protein mixtures (intact *cells*, *tissues*, or *purified protein* fractions) are solublised in a denaturing solution containing *nonionic* detergent(s), urea, a reducing agent (e.g., 2-*mercatoethanol*, *dithiothreitol*, etc.), and carrier *ampholytes* and applied to first-dimension *cylindrical* tube gels containing an appropriate mixture of carrier *ampholytes*. In an alternative procedure, commonly known as immobilised pH gradient (IPG) *electrophoresis*, solublised samples may be applied to *ultrathin* precast *polyacrylamide* gel strips containing a performed immobilised pH gradient. Sufficient voltage is applied to the tube gels or strips to induce the individual proteins to migrate to their respective pI values as determined by the protein's content of specific acidic and basic amino acid residues. The *polyacrylamide* gels are then extruded from the glass tubes using a transfer buffer containing sodium dodecylsulfate (SDS). As a result, the SDS binds to individual proteins and the tube gel is affixed directly onto the top of a second dimension *polyacrylamide* slab gel containing SDS. When voltage is applied, the individual proteins, which are negatively charged as a result of their association with SDS, migrate from the cylindrical tube gels (gel strips) and into the slab gel, where they are further separated by molecular sieving according to their size (molecular weight). The resultant *electrophoreto-gram* is a two-dimensional map in which individual proteins (*polypeptides*) are displayed as discrete spots, with the slower migrating, higher molecular weight proteins located toward the top of the gel and the more *acidic proteins* to the left.

The autoradiogram of a typical 2D-PAGE separation of approximately 1000 [^{35}S] methionine-labeled whole-cell *lysate* proteins from cultured rat liver *epithelial* cells over the pH range of 5-7 and *molecular* weight of 14,000-205,000. Actin, (pl 5.7, MW 43,000), a *ubiquitous* structural protein found in a wide variety of cell types, is identified for reference. Although such maps appear extremely complex, the technique is highly *reproducible*, and *routine analysis* of the 2D-PAGE patterns can be performed by super imposing the *photographic* image of one gel over that of another on a light box. For more complex analysis, *sophisticated* computerassisted image analysis programs are available.

TECHNIQUES

First Dimension: Isoelectric Focusing

First-dimension *isoelectric* focusing (IEF) is typically performed in 4.5% *polyacrylamide* tube gels (1 mm × 160 mm) containing 2%

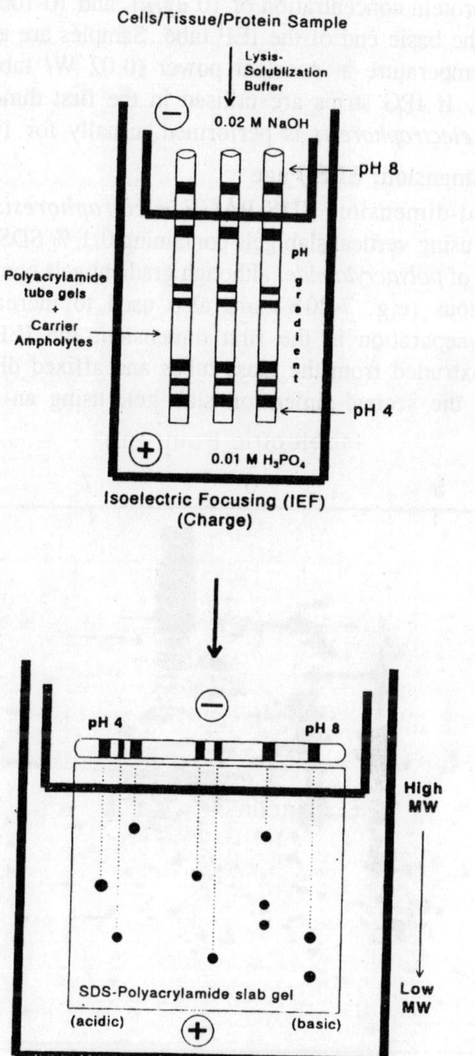

Figure 5.1 : Schematic representation of first-dimension (IEF) and seconddimension (SDS-PAGE) procedures of 2D-PAGE.

carrier ampholytes (1.6% pH range 4-8,0.4% pH 3-10). Protein samples are dissolved in denaturing *lysis* buffer (9.5 M urea, 1 % NP-40, 1 % CHAPS, 10 mM dithiothreitol, and 2% carrier ampholytes) to a final protein concentration of 10 μg/μL and 10-100 p.g of protein loaded at the basic end of the IEF tube. Samples are electrophoresed at room temperature at constant power (0.02 W/ tube) for total of 13.5 kV h. If IPG strips are utilised in the first dimension, flatbed horizontal *electrophoresis* is performed, usually for 100-200 kV-h.

Second Dimension: SDS-Page

Second-dimension SDS-PAGE *electrophoresis* is typically performed using vertical slab gels containing 0.1 % SDS and a constant percentage of *polyacrylamide*, although gradient gels containing variable concentrations (e.g. 7-20%) are also used to increase resolution. Following separation in the first dimension, the IEF gel rods are carefully extruded from the glass tubes and affixed directly onto the surface of the second-dimension slab gel, using an SDScontaining

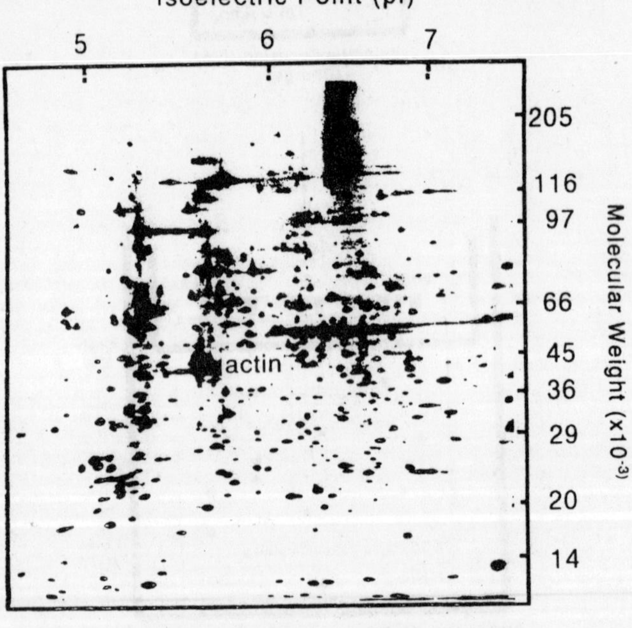

Figure 5.2 : Autoradiograph of [^{35}S]methionine-labeled polypeptides from cultured rat liver epithelial cells obtained by two-dimensional polyacrylamide gel electrophoresis. Polypeptides were separated in the first dimension by IEF in the pH range of 4.5 (left) to 7.5 (right) and molecular weight range of 14 to 205 kDa.

transfer/equilibration buffer and held into place with 1% agarose. IPG strips are treated similarly, although a separate equilibration step of the gel strips in an SDS-containing buffer is necessary before transfer to second-dimension slab gels. Electrophoresis is performed at constant current of 20 mA/gel, using a running buffer of 0.192 M glycine, 25 mM Tris base, and 0.1% SDS. Following electrophoresis (4-5 h), individual polypeptides are routinely visualised using a variety of techniques including Coomassie Blue dye staining, silver staining, and *autoradiography* if cell lysates were metabolically labeled using radioactive amino acids [35S 14C 3H 32p ^{125}1] prior to *electrophoresis*, thereby providing the extremely high degree of sensitivity of 2D-PAGE.

ANCILLARY TECHNIQUES

Practically all analytical techniques applicable for use following 1 D-PAGE can similarly be utilised in 2D-PAGE. Two of the most popular procedures employed in combination with 2D-PAGE are Western blotting and *immunoblot* analysis, using monoclonal and polyclonal antibodies for the identification and analysis of individual proteins. This combination technique has been and continues to be the method of choice to identify specific proteins on the twodimensional maps. Although the vast majority of applications to datc have been for analytical purposes, two new and exciting applications utilising 2D-PAGE are rapidly emerging. The technique is beginning to be used more and more frequently for the "*micropreparative*" isolation of extremely pure polypeptides in sufficient quantities for the generation of both monoclonal and polyclonal antibodies. In addition, recent analytical advances in the area of protein microsequencing have now made it possible to obtain *N-terminal* and internal microsequence information for individual *polypeptide* spots isolated directly from 2D-PAGE gels. This information can be used for the synthesis of appropriate *oligopeptides* for antibody preparation, as well as for the synthesis of appropriate *oligonucleotide* probes for cDNA isolation, gene cloning, and subsequent genetic analysis. These extremely exciting applications provide a valuable link between protein *biochemistry* and molecular biology.

6

Electrophoretic Methods

In *electrophoresis* the charged components to be separated migrate in an electric field at different rates which are governed primarily by the charges and by the strength of the electric field. The charge on a species may be *intrinsic* or produced by an ion bound to the surface of the species.

The motion of particles in a medium is impeded by the frictional resistance, which increases in proportion to the diameter of the particles and to the *viscosity* of the solution. Besides friction, other factors may inhibit the motion, such as ions of opposite charge to that of the particles. The resulting ionic sheath, or '*ion cloud*' is itself associated with a solvation sheath, which likewise impedes the motion. This braking force is the larger, the higher the ionic strength of the buffer, and thus the rate of *migration* decreases with increase in the ionic strength in the electric field.

Electrophoresis may be used for analytical and preparative separation of mixtures of *molecular* or *colloidal* particles. The direction and distance of migration may be important data, and by the new methods developed, especially in recent years, it may be possible to determine the molecular weights and isoelectric points of the components of a mixture.

The technical variants of electrophoresis can be classified in various ways, but for protein work the most useful is according to the field of application: analysis or preparation. There are three main types of each.

Analytical Electrophoresis

(a) Free, frontal *electrophoresis* in a liquid, with a density gradient.

(b) Electrophoresis on porous-structured supports such as paper, cellulose acetate foil, powders.

(c) Electrophoresis in gels, e.g. starch, agar or *polyacrylamide* gels.

Preparative Electrophoresis.

(a) Continuous free *electrophoresis* without a support.

(b) Continuous or zonal *electrophoresis* on paper or on a gel support.

(c) Continuous column *electrophoresis* on a convection-inhibiting, *nonadsorbing*, *granulated* or powder support.

In terms of the technique itself, however, the medium used for the separation is a more *convenient* basis, and in this respect the resolving power of an individual system is an important criterion. Depending on the *medium* and *method* used, for example, *serum* proteins may be separated into different numbers of fractions, as follows:

Type of electrophoresis	Number of fractions
Paper	5-7
Agar gel	8-10
Cellulose acetate foil	9-12
Starch gel	18-20
Polyacrylamide gel	20-50

FREE ELECTROPHORESIS

In free electrophoresis the separation is performed in solution, without a support. The Tiselius *apparatus* is a U-tube containing a *homogeneous* solution of the sample. If an *alkaline* buffer is used, the *protein* molecules migrate towards the anode in the order of their isoelectric points. Detection is based on change in refractive index.

The Tiselius apparatus is now outmoded by *instrumental* development, but is still useful for determination of the *mobilities* of ions in free solution.

A significant advance in support-free *electrophoresis* was the introduction of Hannig's continuous, support-free apparatus. Free-flow *electrophoresis* without a support has the obvious advantage that there is no support to cause loss of certain *macromolecules* by irreversible adsorption; even more importantly, however, the constituents of fibrous proteins (e.g. collagen molecules in solution), *viruses*, *plant* and *animal cell* suspensions, and *haemocytes* can be separated into various fractions.

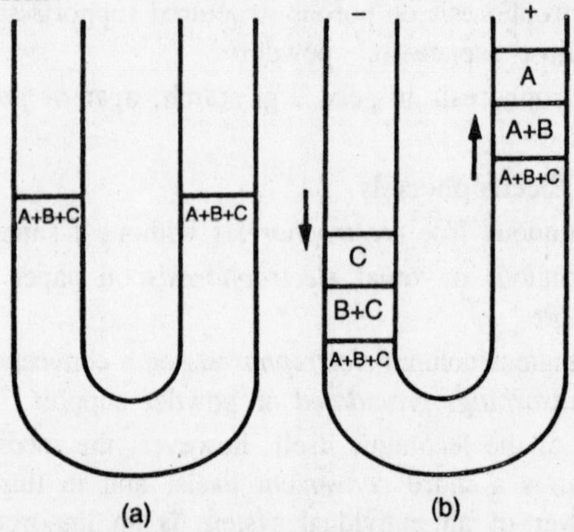

Figure 6.1 : Outline of the Tiselius free electrophoresis process. (a) Before separation. (b) During separation.

The principle of the apparatus is similar to that of the continuous-flow paper electrophoresis apparatus. The separation compartment consists of two vertical cooled glass plates, 0.6 mm apart, the space between them being filled with buffer solution. A solution of the substances to be separated is added continuously to the compartment. The separated fractions are collected in 48 vessels situated at the bottom of the compartment. The nature of the fractions, which may be obtained in preparative amounts, is of particularly great importance in cell research. Hannig *et al.* carried out separations by free-flow *electrophoresis* (without support) on liver homogenates, blood, tumour cells and plant (*spinach*) cell homogenates.

PAPER ELECTROPHORESIS

The spread in use of electrophoresis is largely due to the paper electrophoresis methods developed by Grassmann and Hannig. Although developments in this field in the past 25 years have outstripped this simple method and provided indispensable tools for protein analysis, this method is still useful for medical *diagnostics* and rapid separation of peptides and oligonucleotides.

Because of the pore-size of the support, the separation effect in paper electrophoresis is close to that in free liquid electrophoresis. The spaces between the paper fibres are of such a size that they exert a filtering effect on the particles migrating in the electric field.

During the electrophoresis, the filter paper adsorbs neither the *globular* particles to be separated nor the ions of the buffer used, and the fibres do not dissolve in the buffer.

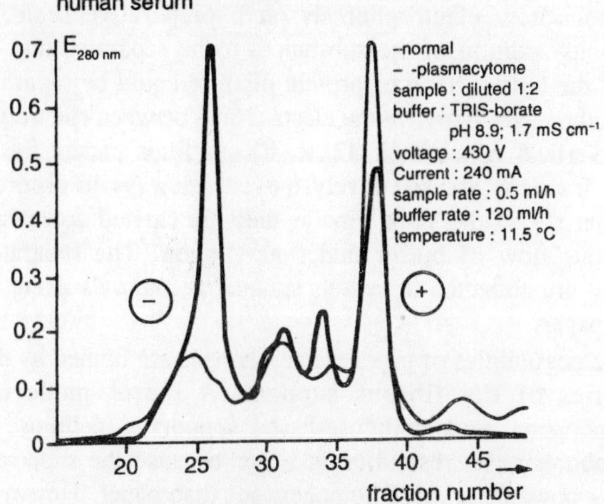

Figure 6.2 : Preparative electrophoresis of human serum proteins with a free-flow electrophoresis apparatus.

The horizontal compartment necessary for paper electrophoresis is cheap and simple, and may be made in the laboratory. The electrodes are immersed in buffer solutions in two *labyrinth*-system compartments, and between them is stretched the support, a filter-paper strip saturated with buffer. The sample is applied in a narrow zone on the filter paper, and in the electric field the component proteins migrate towards the electrode of opposite charge in accordance with their charges. This feature determines the position at which the sample is placed on the strip.

The technique can be conveniently exemplified by the electrophoresis of serum proteins. A 4 × 40 cm paper strip is saturated with buffer solution, and 3-6 μl of serum (0.2-0.4 mg of protein) are placed on it, 4 cm from the cathode end. It is best to use a Michaelis veronal buffer of pH 8.6 and ionic strength 0.1. (For other proteins, the *composition*, *ionic strength* and pH of the buffer are selected in accordance with the properties of the *proteins*.) The *electrophoresis* is performed for 3-14 h at a potential of 110 V with cooling. The *proteins* can be detected on the strip by thermal coagulation and staining with a suitable dye.

Paper electrophoresis can yield reliable values for five basic types of *serum protein*, and indicate whether their amounts correspond to a normal or *pathological* state.

Grassmann and Hannig further employed a much higher potential for the electrophoresis, which necessitated cooling the paper. With the paper vertical, two-dimensional separation became possible; with continuous-flow electrophoresis on a preparative scalé, involving continuous addition of the substances to be separated and continuous flow of the buffer, 10 g of protein mixture could be separated into its basic components daily. In the electric field between electrodes attached to the vertical edges of a 42 × 40 cm filter paper, the individual protein fractions are selectively moved sideways in accordance with their char.r's, at the same time as they are carried downwards by the continuous flow of buffer under gravitation. The separated Protein fractions are collected in vessels situated at intervals along the bottom of the paper.

The possibilities of paper electrophoresis are limited by the *physical* properties of the fibrous support. A more -modern form of electrophoresis on a cellulosé-based support is cellulose thin-layer electrophoresis. Its resolution is better because the support material, *cellulose* powder, is more homogeneous than paper. However, neither two-dimensional analytical *electrophoresis* on filter paper or a cellulose thin layer, nor the combination of electrophoresis with chromatography in the second dimension, has proved able to compete with the newer methods.

CELLULOSE ACETATE FOIL ELECTROPHORESIS

The use of cellulose *acetate foil* for zone electrophoresis has two advantages over paper electrophoresis: (i) greater separating power; *(ii)* greater sensitivity of detection because the foil is transparent.

In principle, the methods used are the same as for paper electrophoresis. As little as 20-50 pg of protein may be applied to the cellulose acetate strip. The limits of detection are 5-1000 ug of separated protein. The electrophoresis takes 4-5 h at 200 V, and 1.5-2 h at 400 V.

With cellulose acetate foil, good separations can be achieved in the electrophoresis of *serum*, *urine*, *cerebrospinal fluid*, *haemoglobin*, *lipoproteins*, *glycoproteins*, *esterases*, *casein*, *whey*, *various plant proteins*, and the proteins of animal meat extracts.

STARCH GEL ELECTROPHORESIS

The intermicellar spaces of colloid-structured gels prepared from hydrolysed starch are small enough to inhibit the electrophoretic migration of large and asymmetric protein molecules to a greater extent than that of smaller, *globular protein molecules*. These opposing effects, the electrophoretic motion as a consequence of the charge, and the sieving effect of the pores of the starch gel, give rise to enhanced separation and resolution.

The high degree of resolution characteristic of starch gel electropho-resis (18-20 fractions may be obtained from a serum protein) depends on the extent of hydrolysis of the starch. The method was developed by Smithies, who has presented a detailed account of the conditions for optimum *hydrolysis* of the starch, and for the electrophoresis.

In starch gel *electrophoresis*, uniformity of the hydrolysed starch is a prerequisite for reproducible separation. Suitable starch can be bought, or made in the laboratory. The latter possibility is an attractive feature, for a cheap basic material displaying constancy over a long period can be obtained by careful processing of a large enough quantity of potato starch. From starch *hydrolysed* with hydrochloric acid dissolved in acetone according to *Smithies*, it is convenient to prepare a 13-14% gel with the appropriate buffer solution.

The electrophoresis may be performed horizontally, and sedimentation of the sample during *electrophoresis* may then be prevented by using a thinner gel (1-2 mm thick). A thicker gel block (6 mm) may be used for vertical *electrophoresis*, and even preparative amounts may then be separated. For example, a vertical gel with a cross-section of 6.5 × 20 mm is suitable for the separation of 50 p1 of serum. The sample can also be soaked up on filter paper, which is then placed in a groove in the gel; the quantity of sample to be used depends on the thickness of the gel.

The field strength of 70 mA necessary for the electrophoresis requires a potential gradient of 6 V/cm, under which conditions an electrophoresis is completed within 6 h. The *Amido Black* 10 B method is generally applicable for staining the gel. Special dyes may be used for specific separated proteins.

Because of its advantages, starch gel electrophoresis has been further developed by several researchers; the work of Scopes is worth particular attention, because of its wide scope. The vertical cooled starch block is prepared with TRIS-borate and TRIS-EDTA buffers of pH 8.4, containing 6% urea.

Starch gel electrophoresis is excellently suited for study of the *polymorphism* of proteins, and for genetic examination: *haemoglobin*, *haptoglobin* and various *isoenzymes* have been separated. The methods of detection of enzymes are listed in the work of Scopes. The basis of the detection is that NAD(P) reduced by the enzyme is capable of reducing Nitrotetrazolium Blue. In the *positive method,* the NAD(P) reduced by the enzyme or the auxiliary enzyme reduces the Nitrotetrazolium Blue in the reaction mixture to yield a purple colour. In the *negative method,* those sites where the enzymes or auxiliary enzymes oxidize NADH remain light, while the colour of the *Tetrazolium Blue* appears where there is no enzyme.

AGAR GEL ELECTROPHORESIS

A gel prepared from agar-agar (composed of *galactan sulphate* and *agarose*) may be utilised for the successful electrophoresis of virtually all proteins. The possibilities and technical details of the method are given in the *monograph* by Wieme.

Galactan sulphate is electrically charged, and by transferring charge to certain proteins it promotes their movement in the direction of the cathode. Accordingly, to obtain clear-cut effects it is best to prepare an 8.5% gel from the agar used for *bacteriological* purposes. The gel is cut into small cubes after cooling, and kept for 3 days in tap-water and then for 3 days in distilled water, the water being changed twice daily. The washed cubes are melted, then stored in a refrigerator until use. Gel prepared from such washed agar has a lower *electroendosmosis* than a gel prepared from the unwashed agar.

The gel layer is generally formed by pouring the melted gel onto a microscope slide, and 50 μl of sample may be added onto the layer by a sample applicator. At a potential gradient of 10 V/cm the *electrophoresis* takes 60 min. Before staining, the separated protein fractions must be fixed by soaking the slide for 4 h in a solution containing 5% *trichloroacetic acid* and 2.5% *formaldehyde*. Wieme provides complete information on preparation of the gel, and the techniques and applications. Agar gel is particularly important because it is the basic material for *immunoelectrophoresis* by various methods.

IMMUNOELECTROPHORESIS

The method developed by Grabar and Williams is a combination of gel electrophoresis and immunodiffusion. In the micromethod introduced by *Scheidegger*, the sample is subjected to electrophoresis in a purified agar gel on a *microscope* slide then a thin strip is cut out of the gel parallel to the direction of the separation, and antiserum is placed in the groove. The gel layer is next incubated in a moist chamber for 24-48 h, during which *immunoprecipitation* occurs between the antibody and antigen molecules.

Numerous variants of the *immunoelectrophoretic* method are given in the literature, and are used primarily for the study of plasma proteins; because of the *immuno*-effects, these form a separate topic of discussion, and are not treated in detail here. Comprehensive accounts of *electroph-oretic* methods based on immuno-reactions, and description of the most, widely recommended procedures, may be found in the books by Backhaus and by Allen, Hill and Stokes.

POLYACRYLAMIDE GEL ELECTROPHORESIS (PAGE)

The unique advantages of polyacrylamide gel electrophoresis (PAGE), namely the simple technique and the extremely high resolution, clearly give this method priority over all previous electrophoretic procedures. PAGE is generally suitable for the separation and analysis of all proteins over the range from polypeptides to ribosomes and viruses. If the buffer composition is appropriately chosen, proteins of neutral, basic or acidic character can be separated with equally good results. As a support, *polyacrylamide* has the outstanding property that it is completely inert, neither changing nor adsorbing the protein molecules. The possibility of concentration also gives rise to special fields of application, such as the direct electrophoresis of very dilute protein solutions.

Most striking is the high resolving power, exemplified by biochemical and clinical laboratory results that could be achieved only by this method, such as the exact '*mapping*' of the basic units of proteins of high molecular weight (e.g. *myosin*, various *enzymes*, *virus proteins*) and the detection of *polymeric* proteins, which otherwise appear *homogeneous*, and their mono-, di-, tri- and *tetrameric* forms. With a differential method, Wright and Mallmann were able to detect 44-50 serum proteins in blood plasma, while with a twodimensional slab method they were able to find 90-95 serum protein components.

An important field of application of PAGE is investigation of the purity of protein preparations. If enzyme preparations are analysed by means of

PAGE, the effectiveness of the purification method may be monitored during purification, and the protein composition of the pure preparation can be analysed.

Electrophoretic separations are governed fundamentally by the physicochemical characteristics of the proteins. Factors influencing the separation are the concentration and steric network of the gel, the composition of the gel and *electrode buffers*, and the magnitude of the electric field applied.

The Basis of PAGE

Polyacrylamide gel is a polymerisation product with a steric network structure, formed from *acrylamide* and N,N'-methylene-bis(acrylamide) (BIS). The *polyacrylamide* chains can connect through their free functional groups with the neighbouring chains, and thus a three-dimensional gel structure results. The *steric network* produced is irregular in form, corresponding to maximum entropy.

The network structure, which is determined by the concentrations of acrylamide and BIS and also by the degree of polymerisation, gives a sieving effect which fundamentally influences the electrophoretic separation.

Polyacrylamide gel has a number of advantages over other gel materials: (a) chemical stability and inertness;

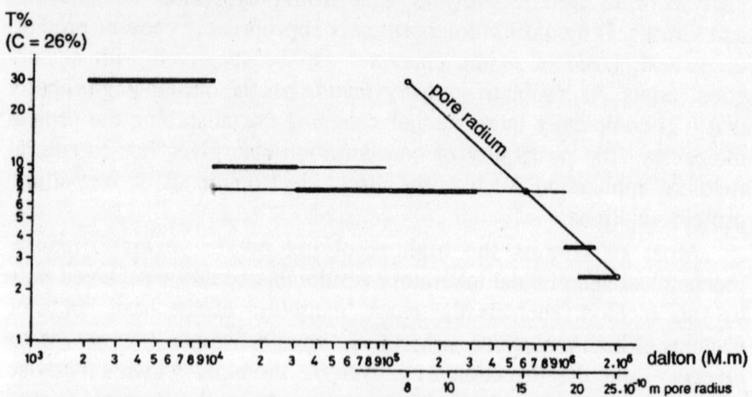

Figure 6.3 : Correlations between the acrylamide concentration of the gel and the pore radius, and the various concentrations and molecular weights of the proteins passing through the pores.

(b) high degree of *transparency*;

(c) pore-size may be selected within wide limits; (d) no adsorption or *electro-osmosis*; (e) insolubility in most solvents.

The rod form of disc electrophoresis developed by *Ornstein* and by Davis uses three different gel compositions. The proteins to be separated are contained in the sample gel, which is large-pored and has a low *acrylamide* concentration. Below this is the collecting or spacer gel, which is likewise large-pored and has a low *acrylamide* concentration, in which the material to be separated is concentrated in the first phase of the *electrophoresis*. Below this in turn is the smaller-pored, more concentrated separating gel, in which the material is separated into its individual fractions.

In the electrophoresis, the ions of the buffer in the gel are the '*leading*' ions, while the ions originating from the buffer in the *electrode* vessel are the '*following*' ions. At the beginning of the separation,

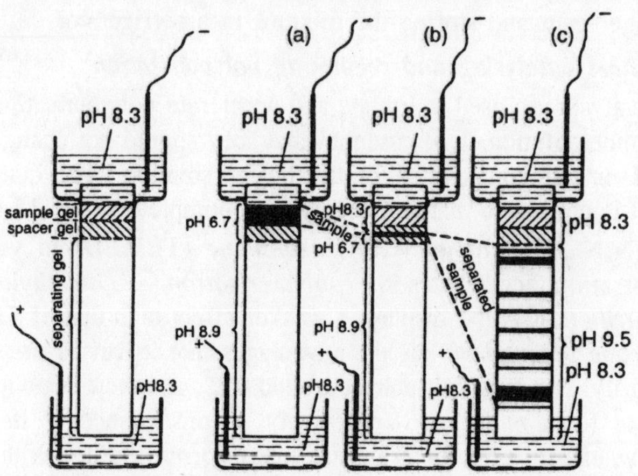

Figure 6.4 : Process of rod electrophoresis according to Ornstein. (a) Start. (b) Protein concentration. (c) Separation of protein

there are leading ions in the sample and spacer gels; the following ions are only in the *electrode* buffer. When the current is switched on, the higher mobility of the leading ions causes them to move in advance of the *proteins* and the following ions in the *spacer gel*, leaving behind them a zone of lower conductivity and higher field strength. This accelerates the proteins and the following ions, so that they migrate at the same rate behind the leading ions and hence accumulate in the spacer gel. When the migrating protein zone reaches the dividing line of the spacer and separating gels, it comes into a buffer of such pH that the rate of the following ions almost attains that of the leading ions and exceeds that of any of the protein molecules; here, therefore, the following ions migrate behind the leading ions, but in front of the proteins.

The Chemical Materials for PAGE

Acrylamide and bis-acrylamide (BIS)

The matrix-forming substances of the polyacrylamide gels are acrylamide (CH_2=CH—CO NH_2) and N,N′-methylene-bis(*acrylamide*) (Cl_2=CH—CO—NH—CH_2—NH—CO—CH=CH_2). Numerous companies market preparations of very high purity, which may be used without purification. If necessary, acrylamide can be purified by precipitation with benzene from an acetone solution, and BIS by *crystallisation* from acetone solution. After the *acrylamide* and BIS solutions have been mixed, any free acrylic acid may be bound by adding Elgalite (1 g/100 ml of solution) or *Amberlite* MB I ion-exchange resin and storing the mixture in a refrigerator.

Substances catalysing and regulating polymerisation

Catalysts are used to initiate and accelerate polymerisation of the acrylamide solution. It is essential that they should not change either the pH or the composition of the buffer solution. The quantity of catalyst is chosen so that gel formation is complete within 15-45 min.

N,N,N′,N′-*Tetramethylethylenediamine* (TEMED) in very low concentration accelerates the *polymerisation*. 3-*Dimethylaminopropionitrile* (DMAPN) displays a weaker effect than that of TEMED. *Ammonium persulphate* has the advantages that it can be prepared in high purity, is relatively stable at around 0°C, and there is no tendency for it to form *molecular oxygen*. As a consequence of its strong oxidative effect, it has the drawback that it forms artefacts with certain of the proteins to be separated, and inactivates some enzymes. This can be avoided in part by removal of the catalyst from the gel by electrophoresis for one hour before application of the sample; alternatively, an antioxidant [*dithiothreitol* (DTT) or 2-*mercaptoethanol*] can be added along with the persulphate before the polymerisation.

It is also customary to use the various catalysts in combination: persulphate + DMAPN, persulphate + TEMED, riboflavin + TEMED.

In acidic medium the polymerisation proceeds very slowly, e.g. in 8-10 h at pH 2.9, and Jordan and Raymond therefore developed a special catalyst for the preparation of an acidic gel: 0.1% *ascorbic acid*, 0.0025% *iron*(II) *sulphate* and 0.03% *hydrogen peroxide*. The *peroxide* must be added to the acrylamide solution immediately before it is poured out, and the gel solution must then be cooled to 5-10°C. The *polymerisation* takes place in 3-5 min.

A substance that can regulate the polymerisation is *potassium ferrocyanide*, which is employed on occasions to ensure uniform effect of the catalyst.

Surfactants

Surfactants such as urea and sodium *dodecyl sulphate* (SDS) further extend the possibilities of application of PAGE.

Urea considerably decreases the adsorptive interactions of the *protein monomers* bound to one another; in a urea solution of high concentration, therefore, these polymers dissociate into their sub-units. Urea treatment denatures protein molecules, however, destroying their original properties. Reductants must be added to the gel solution simultaneously with the urea, to prevent formation of disulphide bonds, i.e. the reaggregation of the sub-units, and hence the production of artefacts. 2-*Mercaptoethanol* or *dithioerythritol* (DTE) in a concentration of 1-5% is generally used as reducing agent.

The effect of SDS arises from its enhancement of the *hydrophilic* character of proteins. The size of a protein molecule does not change significantly when SDS is bound to it. During electrophoresis in the presence of SDS, the effect of the charge of the protein *molecules* on the migration rate is not manifested; only the molecular-sieve effect governed by the pore size of the gel is displayed.

Buffer solutions

A basic requirement of the buffer solutions used is that their composition and pH should not alter the chemical and biological properties of the substances to be separated, and the interaction between protein *molecules* should be minimal. During migration of the sample in the gel, the charges of the leading and following ions must have the same sign. It is not permissible for the mixture to be separated to change the pH of the gel appreciably in the course of the separation; it is therefore recommended that the pH of the sample solution should be adjusted to within ± 0.5 pH unit of the pH of the *electrode buffer*, before the sample is applied to the gel.

Richards *et al.* give equations for calculation of the concentrations of the leading and following ions, and also the counter-ions, together with the pH of the gel and the *electrode buffer*.

The necessity for buffers of higher ionic strength, and the lack of any interaction between the protein molecules and the buffer, are two opposing requirements; attempts have been made to solve this problem by the addition of I=;DTA or other special reagents to the buffer.

Analytical Application of PAGE

Apparatus

There are two methods for performing the *electrophoresis*: the

Table 6.1 : Base solutions for polyacrylamide gel (Quantities given to be dissolved and made up to 100 ml)

No. of stock. soin.	Acrylamide, g	BIS, g	Urea, g	Persulphate, g	Riboflavin, g	DMAPN, Ml	TRIS, g	Gly, g	$K_3Fe(CN)_6$, g	Sucrose g
1	30.0	0.8								
2	60.0	1.4								
3	60.0	1.2								
4	10.0	2.5								
5	28.0	0.735	48.0							
6	10.0	2.5	48.0							
7				0.14						
8				0.18						
9				0.28						
10				0.48						
11				1.00						
12				2.8						
13				10.0						
14					0.004					
15						1.6	0.3	1.44		
16									0.03	
17			48.0							
18										40.0

rod and the slab procedures. The various types of apparatus can therefore be divided into two groups. The apparatus for the simple rod method consists of an upper and a lower buffer compartment, connected by the gel rods: gel-filled glass tubes dipping into the lower compartment are fitted into equidistant holes in the bottom of the upper buffer compartment. The *platinum electrodes* are introduced into the buffer compartments.

Many authors have striven to replace these cylindrical gel tubes with square or rectangular cross-section tubes made from glass or '*Plexiglas*', because the gels can then be more accurately evaluated *densitometrically*. Such types of apparatus represent a transition towards the slab types, as gels prepared in this way can be regarded in essence as parts of a cut-up slab.

The slab apparatus types consist of two parts, a cell for the pouring of the gel slab, and the *electrophoresis* apparatus, which is similar to that for horizontal paper *electrophoresis*.

Preparation of the gel

The most important aspect in selection of the gel is the character of the protein or protein mixture to be examined. Attention must be paid to the size of the molecules and their charge, pH-sensitivity and tendency to undergo association. Thus, the buffer and gel used for the separation of proteins which are *predominantly* of low molecular weight are of completely different composition from those used for proteins of high molecular weight. Deciding features are the pore-size of the gel, and the pH and ionic strength of the buffer.

Stock solutions are generally used for preparation of the *polyacrylamide* gel and the electrode buffers.

With the exception of the persulphate solution, which is best prepared fresh at least weekly, the solutions may be stored for even 2-3 months at 4°C, but the auxiliary sucrose solution readily undergoes mould formation, even if kept in the cold.

Blattler *et al.* have reported their detailed studies on the gel composition. Hjerten uses the following formulae to express the proportion of BIS in the gel:

$T=$ g. of acrylamide + g of BIS in 100 ml of solution

$$C = \frac{\text{g of BIS in 100 ml of solution} \times 100}{T}$$

As an example, gel 3 in Table 3.4 (7.5 g of acrylamide and 0.2 g of BIS per 100 ml) would be described as $T= 7.7\%$, $C = 2.6\%$.

Preparation of gel rods

Gel rods may vary in thickness from 0.1 to 10 mm, with a length of 5-20 cm. The most frequent dimensions are a diameter of 5-6 mm and a length of 8-12 cm. The appropriate quantities of the stock solutions are measured into a small suction flask and carefully mixed. Room temperature is generally suitable for this operation, but in certain cases temperatures of 4-10 °C are prescribed. Before addition of the catalysts, the monomer solution is freed from dissolved gases by application of a vacuum. In the preparation of the layered gel by the method of Ornstein, a solution of the separating gel is first poured into the tubes, up to the appropriate height, distilled water is layered to a height of 0.5-1 cm on the still liquid monomer solution, care

Table 6.2 : Buffer solutions for PAGE.

No.	pH	TRIS, g	Gly, g	β-Ala, g	DL-Val, g	TEMED, ml	12 *M* HCl, ml	1*M* HCl, ml	1NH$_3$PO$_4$, ml	H$_3$BO$_3$, g
A. GEL BUFFERS										
Quantities given are to be diluted with distilled water to 100 ml.										
19	8.9	36.6				0.23		48		
20	6.9	5.7							25.6	
21	7.5	6.85				0.46		48		
22	4.3					4.0				
23	2.9	1.15								
24	5.5	4.95				0.46			30	
25	5.9									
26	8.9	12	3							
27	6.7	1.2	0.3						*	
28	6.7	0.75					0.4			
29	8.9	4.6					0.4			
30	2.9					0.4				
31	2.9	1.45								
32	9.2	2.23								0.37
33	8.9	36.6					4.8			
34	6.7	5.98					4.8			
35	6.7					0.46				

No.	pH	TRIS, g	Gly, g	β-Ala, g	DL-Val, g	TEMED, ml	12 *M* HCl, ml	1*M* HCl, ml	1NH_3PO_4, ml	H_3BO_3, g
B. ELECTRODE BUFFERS										
Quantities given are to be diluted with distilled water to 1000 ml										
36	8.3	6	28.8							
37	8.3	215.5								110
38	7.0	1								
39	4.5			31.2						
40	4.0		28.1							
41a	4.0				35.13					
41b	4.0		22.53							
42	4.0		27.78							
43	9.0	22.31								0.73

* =adjust pH with necessary amount

CH_3COOH, ml	Diethyl-barbit-uric acid, g	Citric acid, g	1 *M* KOH, ml	Na_2 EDTA, g	Urea, g	Note
						for separating gel
						for spacer gel
						for separating gel
17.3			48			for separating gel
53.25			12			for separating gel
						for spacer gel
2.95			48			for spacer gel
						for separating gel
						for sample gel
						for sample gel
						for separating gel
*			4.2			for separating gel
		*				for separating gel
				0.19		for separating gel
					48	for separating gel
					48	for separating gel
2.87			48			for separating gel

CH_3COOH, ml	Diethyl-barbit--uric acid, g	Citric acid, g	1 *M* KOH, ml	Na_2 EDTA, g	Urea, g	Note
						to be diluted 10×
				37		to be diluted 20×
	5.52					
8.0						to be diluted 10×
3.06						to be diluted 10×
* (upper)						anode buffer
*						cathode buffer (bottom)
		*				electrode buffer for No. 32 buffer
				1.92		to be diluted 2×

being taken that no mixing occurs, and the monomer solution is left to polymerise in natural, fluorescent or ultraviolet light. In the formation of neutral or alkaline gels, the *polymerisation* is generally complete within 1020 min. In acidic solution the process is very slow with the customary catalysts; to shorten the *polymerisation* time, it is necessary to use either a larger quantity of catalyst or the catalyst of *Jordan* and *Raymond*. The completion of the polymerisation is indicated by a sharp *demarcation* line between the water and the gel. The water is then sucked off, the residual traces being absorbed with a strip of filter paper. The spacer gel is then layered onto the separating gel in a similar way, and finally the sample gel is poured onto the spacer gel. In the last stage, however, care must be taken that the solution of sample gel added should contain an exactly known amount of protein.

Depending on the diameter of the tube, this may be 50200 μg. The experiments of Hjerten indicate that the sample and spacer gels may be dispensed with if the protein solution to be separated is diluted with buffer, or if 0.01-0.02 ml of distilled water is layered onto the sample applied to the separating gel. Hence, since the potential gradient increases as the sample proceeds downwards in the gel rod, the various protein zones are sharply differentiated. Michl and Pastuszyn state that a 1-2% agarose solution is also suitable for the

Table 6.3 : Quantities of BIS to be used for the preparation of polyacrylamide gels of various concentrations.

Acrylamide, g/100 ml	BIS, mg/100 ml	C%
3.75	80-330	2.1-8.8
5.0	65-290	1.3-5.8
7.5	46-210	0.6-2.8
15.0	30- 88	0.2-0.6

preparation of a spacer layer. It is likewise possible to employ only a separating layer if the sample, mixed in 1 : 1 ratio with 20% sucrose solution, is layered under the electrode buffer onto the surface of the gel by means of a micropipette or a microsyringe. It should be ensured that the sample solution does not mix with the electrode buffer. Narayan *et al.* report that the *omission* of the sample and spacer gels presents no drawbacks if a sucrose solution is used; indeed, the preparation of the sample gel may even be a source of *inaccuracies*.

A concentration-gradient gel will give a better separation (on the basis of the sizes of the protein molecules) than will a separating gel with homogeneous concentration. A continuous *concentration*-gradient can be prepared with a gradient mixer fitted with two reservoirs containing solutions with different acrylamide concentrations.

Preparation of gel slabs

The preparation of gels suitable for *electrophoresis* by the slab technique varies to a certain extent, depending on whether the electropho-resis will be performed vertically or *horizontally*.

During vertical electrophoresis, the gel slabs remain between the two glass plates used for pouring of the gels; otherwise, the same general rules hold for their preparation as in the case of gel rods. Grooves may be made by various means in the upper part of the gel slab, for application of the sample.

In a horizontal gel slab, the sample groove is either produced by the mould used in preparation of the slab, or cut out of the polymerised gel with a special tool. Only the separating gels are usually prepared in this way.

Depending on the thickness of the gel, 10-40 ,ug of protein in 5-20 pl of solution is a reasonable amount to apply to one sample site; it is mixed in 1 : 1 ratio with 20% sucrose or 8 M urea solution.

In the event of a protein consisting of very many components, even 200 pg may be applied to one site.

The slab method has certain advantages over the rod gel technique:

(i) many samples may be run in parallel;

(ii) the Joule heat is more easily dissipated;

(iii) a two-dimensional variant is possible;

(iv) densitometric evaluation is more accurate, and photographic documentation is more successful;

(v) the slabs can be employed more easily for an autoradiographic method.

Preparation of the sample and its application to the gel

The pH and electrolyte concentration of the sample solution should if possible be the same as those of the buffer of the spacer gel. Before the sample is applied to the gel, the concentration of the protein solution to be exathined must be determined. The sample solution should not mix with the electrode buffer, and the protein should enter the gel quantitatively. Depending on the presumed number of components in the sample, 50-200 μg of protein are applied to one site. A single protein (e.g. *albumin*) can easily be detected even in a quantity of only 0.1 g. In the case of a sample containing 20-30 or even more components, the examination is begun with gel rods with comparatively large diameters, and 200-300 ,ug of protein are applied. Biological fluids (*serum*, *cerebrospinal fluid*, etc.) can be applied directly to the gel.

Serum is best prepared by the following method: 0.1 ml of electrode buffer and 0.2 ml of 40% sucrose solution are added to 0.1 ml of serum. A little 0.1 Bromophenol Blue solution (in a 9: 1 mixture of methanol and acetic acid) is previously added to the buffer. As a marker, this indicates the advance of the front during the *electrophoresis*. (For an acidic gel the indicator dye may be basic *fuchsin*.) Then 10-20 μl of the *serum* mixture is layered onto each sample site. Solutions with high ionic strengths (proteins extracted with salt solution of high molarity) cannot be applied directly to the gel; before electrophoresis, it is essential to dialyse them against the *diluted buffer*.

Instead of *dialysis*, which requires a long time, gel-filtration on a 9.1 ml Sephadex G-25 gel column (Prepacked Disposable Column PD-10, *Pharmacia*) is suitable for the deionisation of solutions of low volume.

The electrophoresis

After application of the sample, the electrophoresis may begin. In *alkaline* buffers the protein migrates towards the *anode*; with vertical gels, therefore, the *cathode* is located at the top. The d.c. supply source should be such that the potential and the current strength can be constantly controlled. When gel rods are to be used analytically, it is advisable to concentrate the proteins with a current strength of 1 mA per gel for 10-15 min, and then to increase the current strength to 2-5 mA per gel. The gel slabs used horizontally generally have larger surface areas than the vertical ones, and a higher potential is therefore necessary for the *electrophoresis*. The horizontal gel slab can be placed on a cooling surface, to conduct away the Joule heat produced during the electrophoresis. A different potential and current strength are necessary for every buffer; for the Ornstein TRIS-glycine buffer, a potential of 250300 V is appropriate. If the potential is less than the optimum, the separation of the zones will not be sharp, while if it is higher than the optimum, the Joule heat produced will impair the separation. The problem of cooling can also be solved by circulating buffer at 4°C through the electrode compartments. It is recommended that an apparatus not fitted with a circulatory cooling system should be used in a cold room ora refrigerator. The electrophoresis must be stopped when the indicator dye has progressed through the gel and has approached to within 0.5-1 cm of its edge. After the voltage supply has been switched off, the gels must be transferred to the fixing or staining solution without delay, to avoid diffusion of the separated substances. The electrode buffers may be used several times, but only with the same electrodes.

The quality of a separation depends on the amount of protein applied, and also on the distance of electrophoretic migration. A good separation can be achieved within an upper limit of 30,ug for each component; the lower limit is governed by the intensity of the staining dye employed.

The gel rods may be removed rapidly from the tubes with the aid of an injection syringe needle. The method of removing gel slabs depends on the construction of the apparatus employed.

Fixing, staining and washing of the gel

Before staining, it is generally necessary to fix the separated proteins, not only to prevent diffusion of the proteins, but also to prepare it for the dye solution. The composition of the fixing solution

Table 6.3 : Specific staining.

Protein	Dye
Haemoglobin, myoglobin	Benzidine, H_2O_2
Haemoglobin, myoglobin	Benzidine, H_2O_2
Haptoglobin	Dimethoxybenzidine, H_2O_2
Coeruloplasmin	o-Dianisidine
Lipoprotein	Sudan Black
Lipoprotein	Sudan Black
Lipoprotein	I_2 in glacial acetic acid
Glycoproteins	Fuchsin-sulphite
Glycoproteins	Periodic acid
Glycoproteins	Alcian Blue
Glycoproteins	PAS reagent
Proteoglycans, glycosaminoglycans	Toluidine Blue
Phosphoproteins	Ammonium molybdate
Ribonucleic acid	Lanthanum acetate, Acridine Orange
Ribonucleic acid	Lanthanum acetate, Cyanine, chrome alum
Ribonucleic acid	Methylene Blue
Desoxyribonucleic acid (native)	Methyl Green, Crystal Violet
Desoxyribonucleic acid (denatured)	Pyronine-B

may vary, depending on the staining method. After fixation, the protein zones are made visible by staining, and can then be evaluated from the intensity of staining.

The dyes most frequently used are *Amido Black* and Coomassie Blue. In a study of the binding of dyes to proteins, Fischbein demonstrated that the most commonly used dye, Coomassie BB, is bound to different extents to the individual proteins. If Coomassie BB R-250 is used, the lower detection limit for the staining corresponds to less than 1 μg of protein. The quality of staining is affected by the distance covered by the protein during electrophoresis, and by other factors.

The fixing, staining and destaining of the gels can be carried out simply in test-tubes in the case of *gel rods*, and in a glass or plastic bath in the case of slabs. Destaining of the dye may be accelerated by electrophoresis or by diffusion. One of the oldest methods for *electric* destaining of the dye is that of Schwabe, in whose apparatus the dye not bound to protein can be washed out from the gel rods within 6-8 min. *Maurer's apparatus* is based on a similar principle.

Specific staining

Complex proteins may be detected with staining procedures in which the dyes bind to the non-protein component.

Evaluation of Electropherograms

Direct evaluation in ultraviolet light

If the electrophoresis was done in *quartz tubes* or cells, then it can be evaluated immediately after the completion of the separation, by examination with *ultraviolet light*, without staining. Radola and Delincee claim that quantitative measurement is more reliable for slabs than rods. The individual proteins have different aromatic amino-acid contents, so each needs its own calibration curve.

Fluorometric evaluation

This method is based on staining of the protein with a fluorescent reagent before or after separation; its sensitivity is similar to that of direct ultraviolet evaluation.

Evaluation of stained electropherograms

Quantitative evaluation of stained gels is very difficult, and in general only relative differences can be measured. There are two reasons for this: (i) the same amounts of different proteins give different absorbances because they bind to different extents with a given dye; *(ii)* because of the structural properties of the protein molecules, the different dyes are bound in different ways to a given *protein molecule*.

The stained fractions can be evaluated quantitatively only on the basis of a calibration curve prepared for the same protein. Amido Black gives unstable products and the intensities of the protein fractions must be measured within a short time. Coomassie BB is more sensitive, but the background staining is more intense.

The most common errors in the quantitative densitometry are:

(i) the separation of the peaks is not perfect;

(ii) the detector response is not a linear function of the light-absorption;

(iii) most integrators are inaccurate, particularly for small areas. Only fast *electronic integration* meets requirements; mechanical integration is not satisfactory.

Instead of *densitometry*, *elution* and *photometric* measurement of the dye can be used. Fenner *et al.* cut out the stained strips accurately, extracted them by shaking at room temperature overnight with a standard amount (4-20 ml) of 25% *aqueous pyridine* solution, and measured the absorbance of the extract at 650 nm. In the *photometric* method of Kahn and Rubin, 0.8 mm thick gels, the protein content of which is in the range 0.05-2.75 μg, are measured at 590 nm: the cut-out pieces of gel are smeared onto the walls of the *photometer cell*, which is then filled with 8% *acetic acid*.

In our own work the integrated *densitometric* values and the results

Table 6.4 : Correlation of values obtained by densitometry and by elution.

Protein	Regression line 95%	95% confidence band	Correlation coeff.
Actin	$y = 0.05x + 0.01$	0.0132	0.9958
Albumin	$y = 0.08x + 0.02$	0.0328	0.9845
CPK	$y = 0.016x + 0.14$	0.0384	0.9972
F_2-Histone	$y = 0.06x + 0.03$	0.0259	0.9380
Myosin	$y = 0.05x + 0.03$	0.0252	0.9246
Lysozyme	$y = 0.04x + 0.04$	0.0282	0.9712
Pyruvate kinase	$y = 0.04x + 0.05$	0.0456	0.9476
Troponin	$y = 0.07x + 0.05$	0.0427	0.8914

obtained by *pyridine* elution have been compared for 1-40 tg of purified preparations applied on a gel slab 3 mm thick and subjected to electrophoresis in the presence of SDS. The integration values varied from protein to protein, but were dependent on concentration. Correlation of the *densitometric* and elution results for individual proteins gave the regression equations.

Rodbard and Chrambach developed a procedure for testing the homogeneity and identity of the spots obtained. The method of Hsieh and Anderson may be used for quantitative measurement. A known amount of lysozyme (as internal standard) is mixed with the protein to be determined. After the electrophoresis, staining is performed, the gel is subjected to *densitometry*, and the areas given by the protein are determined by *planimetry*, *triangulation* or *electronic* integration. Within certain limits, the integrated areas are directly proportional to the amount of protein. Since the proteins are stained to different extents, only an '*empirical weight*' of the unknown protein can be determined (using Coomassie staining). The Fast Green staining method of Gorovsky can give the '*true*' quantities of protein.

Fractionation of the gel

After separation, it is frequently necessary to identify or recover the fractions, to measure various activities, etc. This may entail cutting-up of the gel. Most equipment for this operates on the principle of the egg-slicer or the guillotine. There are separate procedures for the fractionation of gel rods and slabs. A histological microtome can be adapted for high-accuracy gel fractionation. With the sharp knife of the microtome, soft and frozen gels alike can readily be cut into 0.5-1 mm segments.

Preservation and storage of gels

It is customary to store electropherograms for a longish period. Gel rods may be preserved in test-tubes and slabs in polyethylene bags, with 7% acetic acid, without any change in the intensity of the staining over a very long time. Larger or thicker gel slabs may be preserved by dehydration. For this, it is recommended that the slabs, cut into thin strips, be dried first in 90% ethanol, 5% 2-*propanediol*, then placed between two *cellulose* sheets and *dehydrated* further with silica gel.

Interpretation of one-dimensional electropherograms

It is well known that the number of fractions obtained by means of a single separation is not always a faithful reflection of the actual number of different protein components in the sample. For various reasons, a given protein or enzyme may even appear in the form of several bands. In some cases there is a real difference in the sizes of isomeric molecules, but the cause of the isomerism may be a difference in charge, or a lower degree of ligand binding, and the conformations of the fractions may also differ. It may happen that there is not a real *heterogeneity* of the molecule or fraction, the apparent difference resulting from the means of separation. There are a number of possible reasons for this:

(i) interaction between buffer and protein;

(ii) protein-protein interactions;

(iii) partial dissociation of the given protein polymer during the separation;

(iv) partial aggregation of the protein;

(v) linkage or interaction between the conformational isomers.

To give a correct picture, therefore, a one-dimensional separation is never enough; a number of different methods must be employed to ensure that the *heterogeneity* of the preparation is possible and real. Through assessment of the various factors (isomerism, conformational interactions, ligand binding, molecular weight, sub-unit composition), the combined methods give overall and supplementary information for the evaluation of electropherograms.

Preparative Methods of Recovering Protein Fractions

Separated fractions may be obtained either by dissolution from the cut-up gel, or by continuous elution during electrophoresis. Cutting up the gel is not a satisfactory method for recovery of larger amounts of protein, and continuous elution is better. Numerous instruments suitable for this are described in the literature and commercial prospectuses. The compositions of special buffer; and gels necessary for preparative separation are described by Jovin *et al.*, Gordon and Louis, and Jacobson and Lodish. Very many advantages are displayed by the elution method of Hjerten *et al.*; further development of this has permitted the separation of 1 g of a protein sample containing several components, the resolving power approaching that of analytical methods.

Methods involving elution with a constant flow of buffer have in common a high degree of dilution of the low-mobility components, because of the slowness of the diffusion. This effect can be diminished by reducing the length of the separating gel. Another possibility is manually or automatically controlled discontinuous elution. Dilution of the eluate was successfully avoided by van Jaarsweld *et al.* with a simple instrumental modification. The apparatus of Boyde and Remtulla deserves special mention. This can be operated continuously for 10 days, and is capable of separating 1 g of protein within 2 h.

Special Applications of PAGE

Electrophoresis in divided gel

Clarke introduced a two-sample method of electrophoresis in gel rods for the identification of proteins. A strip cut from thin waterproof paper or plastic is pressed'onto the surface of the gel.. dividing it into two parts. The two protein samples to be compared are placed one on each half of the divided surface, and then subjected to electrophoresis side by side. In the '*split gel*' method of Leboy *et al.*, the gel rod is divided into two parts by a small separating wall. It is also possible to apply the protein sample by first *soaking* the protein solution onto a little filter-paper and drying this (if the protein can withstand drying).

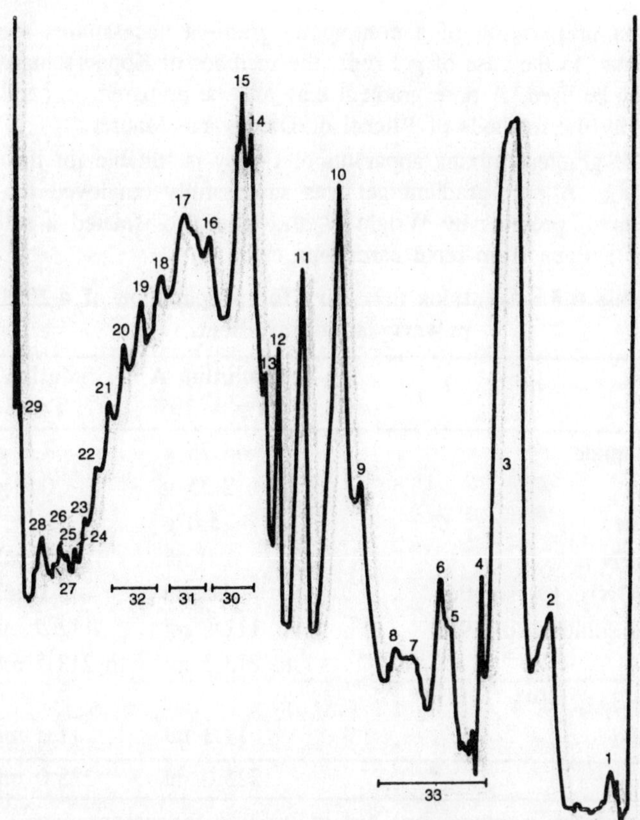

Figure 6.5 : Densitogram of fractions of human serum proteins separated in a stepped gradient polyacrylamide gel slab. Numbering of the proteins: 1, prealbumin; 2, acid α_1-glycoprotein; 3, albumin; 4, α_1-antitrypsin; 5-7, Gc-globulins; 8, coeruloplasmin; 9, haematopexin; 10, transferrin; 11-13, haptoglobin polymers of 2-1 genetic type; 14, β-glycoprotein; 15, α_2-macroglobulin; 1618, haptoglobin polymers of 2-1 type; 19, β-lipoprotein; 20-29, unknowns; 30, γA-globulin; 31, γG-globulin; 32, γM-globulin; 33, α_1-lipoprotein.

Gradient PAGE

With the aim of attaining increased resolution, Slater and subsequently Margolis and Kenrick prepared gels with increasing *acrylamide* concentrations. The concentration gradient may be stepwise or continuous.

Preparation of a stepwise gradient is the simpler, and does not require special apparatus. Allen's procedure can be employed for both rods and slabs, on this basis. We have separated serum proteins on such a gel.

The preparation of a continuous gradient necessitates special apparatus. In the case of gel rods, the methods of Kopperschlager *et al.* may be used. A pore gradient may also be prepared, in capillary tubes, by the methods of Riichel or Dames and Maurer.

The gradient mixing apparatus of Foissy is suitable for the slab technique. A pore-gradient gel was successfully employed for the isolation of proteins by Wright *et al.*, who also isolated a soluble tumour antigen from renal carcinoma cells.

Table 6.5 : Solutions necessary for preparation of 4-20% polyacrylamide gradient.

	Solution A T== 20%	Solution B T= 2
Acrylamide	42.75 g	4.3 g
BIS	2.25 g	0.2 g
Sucrose	25.0 g	
DMAPN	1.0 ml	1.0 ml
$K_3Fe(CN)_6$, 1% solution	4.5 ml	3.0 ml
Buffer solution, pH 9.2*	112.5 ml	112.5 ml
H2^0	to 213.7 ml	to 213.7 ml
$(NH_4)_2S_2O_8$, 10% solution	11.3 ml	11.3 ml
	225.0 ml	225.0 ml

Numerous compositions are in use for the preparation of an acrylamide concentration gradient gel. Those used by Margolis and Kenrick.

The pore-gradient gel of Lorentz contains SDS, which permits an accurate molecular weight determination.

Application of surfactants in PAGE for molecular weight determination

The applications of PAGE are considerably extended by the fact that the molecular weights of proteins treated with SDS can be determined rapidly, with a simple technique, and with an accuracy satisfying practical needs, in *polyacrylamide* gel prepared with an SDS-containing buffer. Treatment with SDS converts proteins into 'random' chains of identical charge, as a consequence of which the electrophoretic migration of the SDS-protein complex is proportional to the molecular weight. However, the migration distance is proportional to the molecular weight only if variation in the hydrodynamic forms and charge/mass ratios of the *macromolecules* is correlated. These conditions are met if the different proteins are bound to the same extent to SDS. In practice, a great many proteins satisfy this requirement, but it must always be ensured that the protein in question is not an exception.

In the standard methods, the proteins are incubated with SDS,

Table 6.6 : SDS-polyacrylamide gel compositions.

No.	Gel pH	Composition of gel buffer		Dilution of gel buffer in gel	Composition of electrode buffer
1	7.1	NaH_2PO, H_2O	7.8 g	1:1	gel buffer: water, 1 : 1
		$Na_2HPO_4 \cdot 2H_2O$	38.6 g		
		SDS	2.0 g		
		H_2O	to 1000 ml		
2	8.4	TRIS	12.0 g	1:4	gel buffer: water, 1 : 4
		Glycine	58.0 g		
		SDS	1,0 g		
		H_2O	to 500 ml		
3	8.2	TRIS	24.22 g	1:2	gel buffer: water, 1 : 2
		SDS	2.0 g		
		AcOH	to pH 8.2		
		H_2O	to 100 ml		
4	8.8	TRIS	32.6 g	1:8	TRIS 3.0
		SDS	0.8 g		Glycine 14.4 g
		6N HCI	to pH 8.8		SDS 1.0 g
		H_2O	to 100 ml		
5	8.96	IN HCI	18.72 ml	1:4	Upper buffer:
					Lower buffer:
		Ammendiol	4.63 g		Ammendiol 4.25 g
					Ammendiol 6.57 g
		H_2O	to 100 ml		Glycine 3.0 g
					1 M HCI 50.0 ml
		(Separating gel)			H_2O to 1000 ml
					H_2O to 1000 MI

mercaptoethanol being added to protect the SH group. Certain authors use dithiothreitol in place of mercaptoethanol.

The general conditions for the gel preparation are as follows:

minimum SDS concentration	0.4%
minimum ionic strength	0.005
time necessary for SDS-protein binding at 100 °C	1-5 min
T	5-15%
C	2-5%

The combination of SDS and PAGE for molecular weight determination was first used by Shapiro *et al.* and by Weber and Osborn. For the determination of *molecular* weights with PAGE,

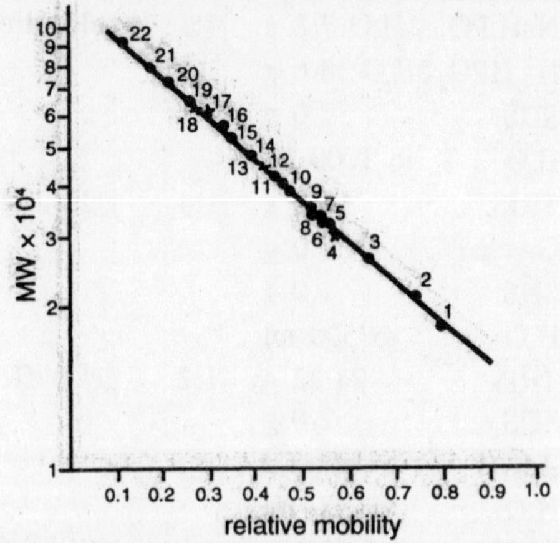

Figure 6.6 : Standards suitable for determination of the molecular weights of proteins separated by SDS-PAGE.1, myoglobin; 2, myokinase; 3, :riose phosphate isomerase; 4, experimental sample; 5, phosphoglycerate mutase; 6, a-glycerol phosphate dehydrogenase; 7, lactate dehydrogenase; 8, experimental sample; 9, glyceraldehyde-3-phosphate dehydrogenase; 10, aldolase; 11, creatine kinase; 12, enolase; 13, experimental sample; 14, phosphoglycerate kinase; 15, phosphoglucose isomerase; 16, pyruvate kinase; 17, 18, experimental samples; 19, adenosine deaminase; 20, phosphofructose kinase; 21, phosphorylase b kinase; 22, phosphorylase b.

standards of known molecular weight are necessary. These must be dissolved in a concentr-ation of 1 mg/ml in the SDS-containing buffer.

About 5 μg each of the standard *proteins* and the *material* under examination are applied to the gel. For complex protein solutions, the opt.mum quantity of total protein and of the individual protein in question must be established empirically. After electrophoresis, the length of the gel and the migration distance of the marker dye are measured. After staining, the migration distances of the standards and the unknown proteins are measured, together with the length of the gel after staining. The relative mobilities, R_1, of the proteins can be established on the basis of the formula:

$$R_f = \frac{\text{protein migration distance}}{\text{gel length after staining}} \times \frac{\text{gel length before staining}}{\text{marker migration distance}}$$

If the logarithms of the molecular weights of the standards are *plotted* as a function of the R_f values, a linear correlation is obtained. After the R_f values of the proteins of unknown *molecular weight* have been determined, their molecular weights can be read off the plot.

Some workers have found a linear correlation between the relative migration distance and the logarithm of the molecular weight. According to Bryan, however, the correlation between the *retardation* coefficient, K_R, and the molecular weight is not linear in the case of polymeric proteins. The deviation can be eliminated by the use of the Ferguson plot (log R_f vs T).

In the absence of a standard protein, an internal standard suitable for the determination of molecular weight can be prepared from the dansyl derivative of insulin or *lysozyme*, according to *Inouye*; the derivative is mixed with the solution of proteins of unknown molecular weight. After electrophoresis, the *dansylated proteins* are detected in ultraviolet light and marked on the gel, and the whole gel is then stained conventionally. The unknown proteins can also be dansylated, and can then be detected and photographed in ultraviolet light after electrophoresis. Determination of the molecular weight on the dansylated sample becomes simpler with the use of the 'logarithmic sample groove preparer' of Catsimpoolas and Kenney.

Chiga *et al.* compared the utility of the ultracentrifugation and SDS-PAGE techniques for the characterisation of human serum proteins in accordance with their molecular weights. They concluded that PAGE gave the better resolution.

It was found by Williams and Gratzer that SDS-PAGE is reliable for determination of molecular weights above 15 000, but below this value it is no longer accurate. Peptides with molecular weights less than 6000 migrate together in the front irrespective of their molecular weights.

In a study of the *electrophoretic* mobilities of membrane proteins in SD-*Spolyacrylamide gel*, Fessenden-Raden observed that the fatty acids in the preparation alter the migration and staining properties of the protein molecules; this circumstance must therefore be taken into consideration in the separation of samples containing fatty acids.

Lambin *et al.* report that SDS-polyacrylamide gel can be employed in either stepwise or continuous gradient form.

SDS-PAGE is also very suitable for isolation purposes. The SDS may be rapidly removed from the isolated preparation by *electrodialysis*, by the method of Tuszynski and Warren.

Triton X-100 may be used in place of SDS for use with PAGE. A method of molecular weight determination has been described by Hearing *et al.*, which preserves the structure and function of the protein; this makes use of Rodbard and Chrambach's upper buffer A, to which 0.1% Triton X-100 is added. The sample is treated with 1% Triton X-100 solution for several hours, then an equal volume of 20% sucrose solution diluted twofold with the upper buffer containing 1% Triton X-100 is added; the current for the electrophoresis is 2 mA per gel rod. The method is of great advantage for the investigation of intact membrane proteins. There is a linear correlation between the effective molecular radius and the square root of K_R.

Nuclear processing of gels containing radioisotope-labelled proteins

Proteins and nucleic acids can in general be labelled with the *isotopes* ... 1, 32p ^{64}Cu and ^{14}C. The γ-activity can be measured with any γ-detector after fractionation of the gel. A gel containing soft β-emitting isotopes must be homogenised and then transferred into solution. Procedures for the dissolution of the gels and measurement of the radioactivity have been published by Alpers and Glickman, Leon and Bohrer, Paus and Tykva and Votruba.

The positions on gel slabs of *protein molecules* labelled with the *radioisotopes*, ^{59}Fe and ^{32}P can be located by autoradiography. If this requires a long time, the gel and the the film must be protected against dryingout, by being wrapped in polyethylene foil. The sharpness of the spot can be enhanced by performing the *autoradiography* at 0-2 °C.

For the autoradiography of proteins labelled with soft β-emitters, it is advisable first to dry the gel. The technical details of the preparation and the autoradiography are given by Fairbanks *et al.* and Uriel and Lavaille. An effective and very economical method for measurement of ^{14}C and 32P activities has been described by Leinen and Witliff. After electrophoresis, the gel is cut up and dried on Whatman No. 1 paper strips, and the samples are then measured in a *scintillation cocktail*. The extraction method of Bouissou *et al.* may be employed after two-dimensional separation.

Separation of enzymes and isoenzymes by PAGE

The separation of enzymes by PAGE does not generally differ from other PAGE methods. An essential aspect is to keep the electrode buffer and the gels at low temperature (2-4 °C) during the electrophoresis, so that the activities of the heat-sensitive enzymes are not decreased.

For the detection of enzymes and the measurement of their activities, the gel must be homogenised after it has been cut up, so that the proteins can he eluted with buffer from the individual fractions. The eluates may be concentrated by *ultrafiltration* and then *lyophilised*, and the *enzymatic* activities may be measured in the normal way.

The following possibilities are available for the detection of isoenzymes:

(i) the acrylamide gel slab is covered with filter paper or cellulose acetate membrane, and the reaction mixture is poured onto this;

Table 6.7 : Specific enzyme detection.

Enzyme
Cholinesterase
Lactate dehydrogenase
Peroxidase
Cytochrome oxidase
Aldolase
Alkaline phosphatase
Acid phosphatase
Arginase
Alcohol dehydrogenase
Amylase
Hexokinase, glucose phosphate isomerase, phosphoglucomutase
Proteinases
Glycosidases
6-Phosphofructokinase
Superoxide dismutase
Nuclease

(ii) the substrate is dissolved in agar or agarose gel, which is then poured onto filter paper; the agar gel slab obtained after congelation is placed on the polyacrylamide slab;

(iii) the substrate is dissolved in agarose gel and poured onto the acrylamide gel slab that has been subjected to electrophoresis, and the agarose gel congeals on it;

(iv) the gel is placed in the incubation solution.

Incubation is generally carried out at 37°C, for a period determined by the rates of all the processes catalysed by the *isoenzymes* to be separated; that is, the incubation must be continued until all of the isoenzymes have reached' maximum activity.

The correlation between the amount of the product and the activity of the enzyme is generally linear. In that case densitometry and peak-area evaluation can be used. However, *non-linearity* may be caused by diffusion of the reacting components, the temperature and duration of the incubation, and the pH of the buffer employed.

For specific detection of enzymes the corresponding histochemical methods are generally used. It is very important that the diffusion of the reagents should not be impeded, in order for the reaction products to accumulate in the outer gel layers. For the detection of enzymes or isoenzymes, therefore, a gel thicker than 3 mm should not be used if possible.

Immunological applications of PAGE

Acrylamide gel is also suitable for immuno-precipitation investigations. Some of its advantages are that preparation of the gel is simple, there is no endoosmotic effect, and the electrophoretic separation can be effected rapidly. It has the drawback that the *immuno-reaction* requires a comparatively long time.

Of the numerous *immuno-electrophoretic* methods, the *discelectrophoresis* procedure of Louis-Ferdinand and Blatt is particularly noteworthy; in this the *antiserum* is *polymerised* into the *spacer gel*, and the antigen into the sample gel. After *electrophoresis*, the tubes are immediately placed in ice-cold water, to inhibit diffusion of the protein; 10-15 min later, the gel is taken out of the tubes and left to stand overnight in an *acetic acid* solution of *Amido Black*. The result may be read off after destaining. From comparison of control and experimental samples by the method of Makonkawkeyoon and Haque, it may clearly be seen which are the *immunologically* related proteins; these may also be measured quantitatively too. Detection of the *antigens* is also possible after SDS-PAGE, by the method of Olden and Yamada.

Two-dimensional PAGE

By one-dimensional electrophoresis in a *homogeneous gel*, the proteins are separated fundamentally in accordance with their charges, whereas in gradient and SDS gels they are separated in accordance with their sizes and molecular weights. However, these methods are insufficient when the sample contains a number of different molecules of the same charge and size. The reason for this is that, in a one-dimensional *electropherogram*, the proteins with similar physico-chemical properties migrate together, and the bands overlap. In their examination of the methods suitable for the separation of complex protein mixtures, Wright et al. obtained most data with two dimensional PAGE, first in a homogeneous gel, and then a continuous 2-30% concave pore-gradient.

The resolution attainable in practice with one- and two-dimensional PAGE methods is governed by the peak capacity and by the sensitivity of detection. In the one-dimensional method, the various limiting factors strongly reduce the maximum resolving power *theoretically* possible. If a correctly selected two-dimensional method is employed, the ideal *theoretical* plate number of 2000 may be approached.

Consecutive use of the different *electrophoretic* methods ensures good possibilities for the isolation of the components of complex protein mixtures which are otherwise difficult to separate.

Microscale PAGE

In principle, any '*macrosystem*' PAGE apparatus may be adapted for a microprocedure. One of the oldest microcell PAGE instruments is that of Pun and Lombrozo. The extremely simple method developed by Hyden *et al.* permits the analysis of proteins even at a cell level. The gel tube is a capillary 55 mm long and 0.215 mm in diameter, with a volume of 2 μl, into which the gel solution is sucked. Making use of the concentrating effect of PAGE, Burr *et al.* developed a *micromethod* which is also suitable for the *electrophoresis* of very dilute protein solutions. The sample is placed in the sample collector, constructed from the funnel-like end part of an

Table 6.8 : Combined electrophoretic methods.

First dimension	Second dimension
PAGE	*PAGE,* different pore size
	PAGE, different pH and pore size pore gradient electrophoresis
	transverse pore gradient electrophoresis isoelectric focusing in polyacrylamide

	gel immunoelectrophoresis isotachophoresis in polyacrylamide gel
pore gradient electrophoresis	*PAGE* pore gradient electrophoresis transverse pore gradient electrophoresis isoelectric focusing in polyacrylamide gel immunoelectrophoresis isotachophoresis in polyacrylamide gel
cellulose acetate electrophoresis	*PAGE* *SDS-PAGE* pore gradient electrophoresis immunoelectrophoresis
isoelectric focusing in polyacrylamide gel	*PAGE* *SDS-PAGE* pore gradient electrophoresis immunoelectrophoresis
isotachophoresis in polyacrylamide gel	*PAGE* *SDS-PAGE* pore gradient electrophoresis *immunoelectrophoresis*

Eppendorf pipette; its volume may be 0.5-0.6 ml. As a result of the concentrating effect, the protein in the dilute sample solution is concentrated into a very small area of the capillary.

Micro-rod electrophoresis was employed by Ri chel, Neuhoff and then by Dames and Maurer, with and without a pore-gradient gel.

For purposes of *slab-microelectrophoresis*, Been and Rasch modified the vertical apparatus of Reid and Bielski, in which the cell contains a thin, film-like gel slab. In the micro-gel apparatus of Maurer and Dati, the dimensions of the gel slabs are 75 x 18 x 0.75 mm, the volume of sample that can be applied is 1-5,u], and the lower limit to the amount of protein detectable is 0.1 μg. The *slab-microelectrophoresis* method may also be used for the preparation of a stepwise gradient, and it may be utilised for enzymatic *micro-investigations*, and *immuno-precipitation* and radioactive techniques.

The Applications of PAGE in Biochemical Research

The PAGE methods have yielded significant results in *enzymology*, primarily in the field concerning the sub-unit structures of enzymes and the study of changes in them.

Detailed investigations have been made with PAGE of lactate

dehydrogenase and malate dehydrogenase in blood, sperm and tissues, of glyceraldehyde-3-phosphate dehydrogenase, of phosphorylase a and b, of DNA *polymerase*, of *ribonuclease*, of *carbonic acid* hydrase, of *hyaluronidase*, and also of various *esterases*, *phosphatases*, *amylase*, *proteolytic enzymes* and *aldolase*. In protein biochemistry, the results of studies of histones with PAGE are extremely important.

The peptide isolation technique of Wada and Shell is of significance as a biochemical method, as is the fingerprint method of Bray and Brownlee for the identification of peptides. In Houston's procedure, the protein zones cut out of stained gels can be utilised for amino-acid analysis. N-Terminal sequence analysis on nanomole quantities can be carried out by the method of Weiner *et al.*.

The results obtained with PAGE in muscle biochemistry have led to an understanding of the sub-unit structures of structural proteins. From our own work, mention may be made of the qualitative and quantitative SDS-PAGE analysis of the muscle proteins, in which the compositions of the myofibrillar, sarcoplasmatic and mitochondrial fractions of rabbit and porcine muscles were clarified. An *electropherogram* of the *mitochondrial* proteins of the soleus, psoas and *gastrocnemius* muscles of rabbit.

Clinical chemical applications of PAGE

In a detailed treatment of the clinical chemical applications of PAGE, Maurer and Allen described the standardisation of sample collection and preparation. Hoffmeister and Felgenhauer have discussed the possibilities of utilising PAGE for clinical diagnosis.

In clinical applications, the collection and storage of the samples to be examined do not differ from those *customary* in protein analysis. In selection of the type of apparatus, the following aspects must be borne in mind: the slab method is more advantageous for the simultaneous analysis of a large number of samples, and the gel-slabs can easily be combined with other separation techniques, e.g. various types of immunoelectrophoresis, and isoelectric focusing. In the qualitative and quantitative evaluation of serum proteins, knowledge of the haptoglobin type of the individual is essential, for without it only a few peaks can be identified. The *electropherograms* of the serum of individuals of the same age and sex, typified for haptoglobin, may be used as reference. In our work we have determined the quantitative distribution of the individual haptoglobin types, and the relative amounts of the individual components.

Table 6.9 : Distribution of components of haptoglobin types

Type	Components						
	I	II	III	IV	V	VI	VII
Hp 2-2							
R_f	0.02	0.08	0.13	0.18	0.28	0.39	
%	24.2	7.35	11.91	25.98	32.39	8.54	
Hp 2-1							
R_1	0.05	0.09	0.13	0.19	0.29	0.45	0.73
%	1.12	2.58	6.67	13.03	25.16	33.85	22.83
Hp 1-1							
R_f	0.75						

The haptoglobin pattern is imprinted on the protein spectrum, as it were, and conceals certain components if a general protein-staining method is employed. Accordingly, it is more correct if, after parallel samples have been run, some strips are stained for total protein, and the others specifically for *haptoglobin*.

On the basis of a large number of serum analyses, Pastewka *et al.* gave the relative proportions of the individual fractions. For the qualitative analysis of serum proteins, the method of Allen may be made use of with discontinuous gel-pore and buffer-ion concentrations but constant pH.

Isolated lipoproteins were first investigated by Narayan and Kummarow.The PAGE of serum *lipoproteins* was dealt with by Muniz

Applications of PAGE in Food and Feed

Protein Research

The various forms of PAGE have proved extremely suitable for the study of food and feed proteins, both for solution of certain basic questions of research, and for quality control.

In our own work it has been found that for the analysis of the myofibrillar proteins of striated muscle, PAGE in combination with *isoelectric* focusing provides more data than any of the previous methods.

ELECTROPHORESIS WITH POROUS SUPPORTS

Besides the fibrous supports and those with steric networks, porous sorbents find special uses in *electrophoretic* separation, owing to their physical characteristics and also to their technical applications. Sorbents suitable for this purpose may be either hydrophobic or hydrophilic, but a requirement for both types is that they should be able to stabilize a large volume of buffer by virtue of their granular structure or their hydration.

Table 6.10 : Food and feed proteins examined by PAGE

Plant proteins
Lemon, fruit
Potato, tuber
Wheat, seed
Carrot, rhizome
Soya, seed
Bean, soluble proteins of seed
Rice, seed, γ_3 globulin
Aspergillus oryzae, enzymatically lysed
Vicia faba, globulins extracted from seed
Bean, globulins of seed
Potato, proteins of tuber
Soya, 7S and 11S globulins of seed
Proteins of various cereal seeds
Soya, seed proteins (in meat products)
Animal proteins
Tryptic cystine peptides of ovalbumin
Extracted proteins of muscle tissues of various domestic animals
Proteins of meat of various fish
Proteins of hen-egg, chromatographic fractions
Emu egg yolk, separation of apovitellenin 1
Porcine muscle structure proteins

Of the *hydrophobic* substances suitable for the *electrophoresis* of proteins and *peptides*, kieselgel kieselguhr, *aluminium oxide*, *silica gel* controlled-pore glass and Pevikon C-870 have proved to be good supports, as have cellulose, Sephadex and polymerised acrylamide powder amongst the strongly *hydrophilic* substances. It is a further basic requirement for hydrophobic supports that they should have a

Table 6.11 : Electrophoretic separation methods combined with IEF

Methods employed	Materials examined
IEF+PAGE	protein solutions of heterogeneous composition
IEF+SDS-PAGE	non-histone proteins
IEF+SDS-PAGE	soluble proteins
IEF+SDS-PAGE	peptides of membrane proteins
IEF+SDS PAGE	normal and virus-infected cells
IEF+PAGE	plasma proteins
IEF+SDS-PAGE	peptides of myofibrillar muscle proteins, non-histone proteins of muscle cell nuclei

very small but uniform grain-size, for these properties determine the water-binding capacities of the support layers prepared from them.

ISOELECTRIC FOCUSING (IFF)

In contrast with electrophoresis, for which a butler of constant pH is used, in *isoelectric* focusing (IEF) a uniformly increasing pH gradient from the anode to the cathode is created by means of carrier ampholyte molecules with different charges. In the pH gradient the individual proteins migrate (under the action of the electric current) to the point where the pH of the medium corresponds to the isoelectric points of the individual protein molecules, and since at the isoelectric points the net charges of the proteins are equalised, and the proteins are outwardly neutral, they are focused there.

The direct aim of IEF may be the isolation of certain component proteins, or the establishment of their *isoelectric* points; in addition, however, the IEF results may give a guide to the most expedient means of further preparative and analytical processing of the material under examination.

The principle of the method was established by Kolin and by Vesterberg and Svensson who produced the carrier ampholyte preparations from mixtures of various *polyamino-* and *polycarboxylic acids*. The aliphatic ampholytes synthesized by Vesterberg and Svensson dissolve very well in water, the pH varying in the range 3-10. The various commerciallyavailable carrier ampholytes (*Ampholine*, *Bio-Lyte*, *Pharmalyte*, *Servalit*) have narrower pH ranges and allow selection of the product with the most appropriate pH interval for the protein concerned (pH 2.5-5, 4-6.5, 6.5-9, 8-10.5). The ampholytes should have high buffer capacities, low conductivity throughout the entire gradient, low and uniform absorption in the *ultraviolet*, no *toxicity*, zero or minimal interaction with proteins, and no chelation properties.

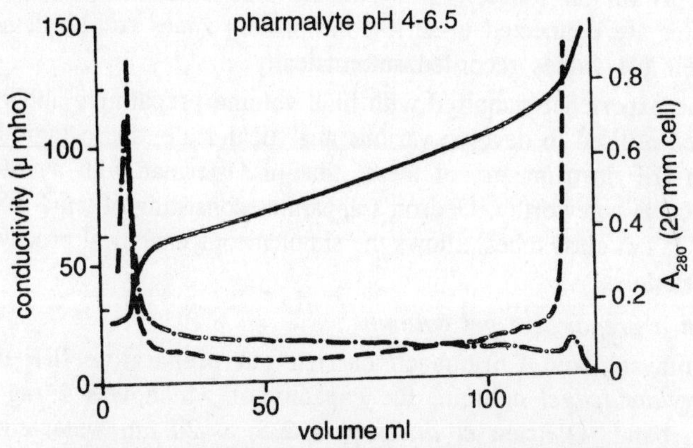

Figure 6.7 : The physical properties of a typical ampholyte.

Like electrophoresis, IEF can be carried out by a column method in a sucrose gradient without a support, or in an acrylamide gel column, by a *polyacrylamide* gel rod method, and also by a layer method in an *acrylamide* gel slab, or in a *granulated* gel layer.

IEF by a Column Method

IEF without a support, in a sucrose gradient

Commercial industrial focusing apparatus consists of *cylindrical* tubes of volume 110 or 440 ml, that can be temperature-controlled both inside and outside. In the original method of Haglund the sample mixed with *ampholyte* of appropriate pH interval is layered into the separating space with a sucrose gradient preparer. At the beginninig of the focusing, a potential of 500 V is generally employed; as the current falls, the potential is increased to 600-1000 V. The separation requires 24-72 h. After completion of the elect rofocusing, the sucrose solution containing the focused proteins is led through an ultraviolet detection-unit into a fraction collector for further processing and analysis. Certain ampholyte preparations strongly absorb light of wavelength 280 nm; if these are used, low protein concentrations cannot be detected by ultraviolet absorption. Before protein determination by the Folin reaction, the fractions must be dialysed for at least 72 h, with repeated exchange of the water. By gel filtration on a Sephadex G-50 column, the carrier ampholytes can be removed more rapidly than by dialysis.

It is essential that the pH should be measured at the temperature at which the focusing was performed. The isoelectric points (pI) of the proteins detected in the fractions may be established on the basis of the pH of the solution. If a suitable ultraviolet monitor and the pH-meter are connected in series, the protein zones can be detected and their pH values recorded automatically.

The experience acquired with high-volume preparative apparatus has been utilised to develop various analytical-scale, sucrose-gradient column IEF instruments; of these, that of Osterman,with a volume of 12 ml, is noteworthy. Godson's apparatus,consisting of ten J-shaped 0.9 × 25 cm glass tubes, allows the simultaneous analytical processing of 10 samples.

IEF on a preparative gel column

Finlayson and Chrambach carried out preparative IEF on a *polyacrylamide gel* column, the capacity of which was 2 mg per protein band. O'Brien *et al.* used a 2.5 × 20 cm water-cooled Sephadex G-15 column for preparative IEF. The separation began with a current of 15 mA (at 800 V), and continued until -it diminished to 2.5 mA.

Analytical IEF in a Polyacrylamide Gel Rod

The method of analytical IEF in polyacrylamide gel was described by Wrigley; the *rod technique* was modified by Fawcett and numerous other authors, and employed for the separation of the most varied substances.

The gels used for rod IEF may be prepared by either chemical or photopolymerisation. Of the large number of gel recipes, the compositions of those most widely used are given below.

Gels prepared by chemical polymerisation:

7.5% gel:

acrylamide solution (T= 30%, C=3.3%)	7.6 ml
ampholyte	0.9 ml
urea	14.4 g
3-dimethylaminopropionitrile (DMAPN)	0.1 ml
$(NH_4)_2S_2O_8$, 10% solution	0.1 ml
distilled water	to 30.0 ml

5% gel:

acrylamide solution (T= 30%, C = 3.3%)	4.5 ml
ampholyte	0.9 ml
urea	14.4 g
3-dimethylaminopropionitrile (DMAPN)	0.1 ml
$(NH_4)_2S_2O_8$, 10% solution	0.1 ml
distilled water	to 30.0 ml

The oxidising effect of the *persulphate* on *proteins* can be eliminated by storing the chemically *polymerised* gels overnight before use, or by subjecting them to prefocusing for about 1 h before addition of the sample

In photo-polymerisation, 25% v/v of 0.1% riboflavin solution may be added to the acrylamide solution instead of *N, N, N', N'*-tetramethyl-ethylenediamine (TEMED) and $(NH,)_2S_2O_8$ catalysts. Photo-polymeri-sation takes place on application of a fluorescent light-source.

The sample must be transferred into the *polymerised gel rods* under a protective solution containing 2% *ampholyte* and 5% *sucrose*, layered 2-5 mm thick on top of the rod. However, very dilute protein samples may be added to the prepared acrylamide solution even before polymerisation. The amount of protein may be 10-20 μg per band.

The anode solution is usually 0.2% *sulphuric acid*, and the *cathode* solution 0.4% *ethanolamine* (or *triethanolamine*). In rod IEF, the initial potential of 200-300 V must gradually be raised to such an extent (generally to 400-500 V) that during the process of separation in 10 gel rods, the current should remain at 40 mA. The IEF is complete when the current no longer decreases; this occurs after about 90-120 min.

The pH gradients of the gel rods may either be determined at 2-5 mm intervals on the surface of the reference gel with a microelectrode, or measured after elution of the cut-up sections of the gel. If the resulting pH values are plotted as a function of distance, the pI values of the individual protein fractions may be read off the curve. Before staining, the protein fractions in the focused gels must be fixed with *trichloroacetic* acid or *trichloroacetic acid+sulphosalicylic acid* solution; the *ampholyte* must then be washed out of the gel before the use of the traditional staining methods. Even without washingout of the *ampholyte*, staining may be performed with a solution of Coomassie BB R-250, either by the method of Reisner *et al*, or with the staining solution of Wellner and Hayes, which contains not only the dye but also the *trichloroacetic acid + sulphosalicylic acid* used for fixing.

Microscale IEF methods have been developed by Bispink and Neuhoff and by Gainer, on the basis of the pattern of microrod electrophoresis. The detection of even 10^{-10} g of protein is possible in gel polymerised in 2, 5 or 10-μl capillaries.

Analytical IEF in a Polyacrylamide Gel Slab

IEF in a *polyacrylamide gel* slab possesses similar advantages to those observed in PAGE, with respect to the gel rod method. The gel slab is generally 1 mm in thickness, while it may vary in length from 9 to 20 cm between the two *electrodes*. If good types of gel pourer and IEF instrument are not available, the gel may be prepared very easily with the apparatus described by Uriel and Berges .

A cooling plate must definitely be inserted below the glass plate supporting the gel, in order to cool the gel slab; 2-20 μl of a solution containing 1-2% protein is applied to the surface of the gel.

With unknown samples it is advisable to begin the IEF with an ampholyte displaying a wide pH interval (3-10). The result gives information regarding the correct choice of *ampholyte* with narrower pH interval, which yields better resolution. which illustrates the results of the IEF of 18 proteins in *polyacrylamide* gel containing *Pharmalyte* with a pH interval of 3-10.

SPreparative IEF on a Polyacrylamide Gel Flat Bed

With a gel slab of preparative dimensions, the electrofocusing of 10-20 mg of protein can be achieved within 6 h, e.g. in a 2-4 mm thick 7.5% gel slab by the method of *Graesslin et al.* The protein zones are located by staining the strips cut off the two edges of the slab, and on this basis the desired zones can be cut out of the slab. The proteins to be further studied can be obtained in preparative amounts from these by elution, and then by concentration of the eluates.

IEF on a Granulated Gel Bed

The most appropriate substances for IEF in granulated gels are Sephadex IEF and Sephadex G-200 SF; these may be used to prepare thin layers of *analytical* dimensions, and also gel beds 5 mm in thickness, suitable for *preparative* separations. IEF carried out on the gel thin layer has no advantages over IEF on the *polyacrylamide* layer, either with regard to operation (preparation of the thin layer and detection are the same) or resolution. However, preparative IEF in a 5-mm thick Sephadex G-200 SF gel bed yields a fractionation that cannot be *approached* either with a *sucrose* gradient in a column system, or separation in a preparative polyacrylamide gel column.

The method has the very great advantage that after IEF in a broad pH interval in a suitable apparatus, the desired fraction or fractions may be selected and removed from the gel bed for further purification with an ampholyte displaying a narrower pH interval.

Combined IEF Methods

IEF is a procedure which may be coupled with other electrophoretic methods, e.g. conventional gel *electrophoresis*; in certain cases, such a *combination* may result in a better resolution.

NON-ISOELECTRIC FOCUSING IN BUFFERS

Three extremely important features of IEF, viz. the instability in time of the pH gradient induced by the carrier *ampholyte*, the existence of a small current in the 'steady state', and the IEF of nucleic acids, led Nguyen *et al.* to examine whether the *multiphase* buffer systems used in PAGE possess the ability to create a stable pH gradient without carrier ampholytes. Their experiments involved the use of *amphoteric* and *non-amphoteric* buffer mixtures, and by means of *theoretical* calculations recipes for very precisely designed buffer mixtures were compiled, by means of which a stable pH gradient, sharp buffer and protein zones, and a constant zone width could be reproduced in the cases examined.

ISOTACHOPHORESIS

In *isotachophoretic* separation, the sample ions to be separated (electrically charged and remaining in the ionic state throughout) exhibit different relative mobilities in an electric field. The ions of the same charge that can be separated by *isotachophoresis* have common counter-ions (e.g. TRIS), the mobilities of which in a constant electric field are smaller than those of the counter-ions of the previously leading electrolyte (e.g. the acetate ion), and larger than those of the ions of identical sign in the terminal electrolyte (e.g. the *glycinate ion*).

At the beginning of the separation, the sample ions are situated among both the leading and the terminal ions. In the separation, they become arranged in accordance with their relative mobilities (approximately in the sequence of their pI values) along the total length of the separation space (chamber, gel layer, column). After equilibrium has been established, the sample ions begin to migrate at the same rate towards the *electrodes* corresponding to their charges. In the course of this migration, at constant current, the electric field varies in a stepwise manner between the *separated zones* (which remain at a constant distance from one another); in this way, the potential is constant in each zone from the front to the start, but increases by jumps from one zone to another.

The separated substances, closely following one another at the same rate in narrow bands, can be kept at greater distances from one another by the addition to the sample of *ampholytes* with pH intervals corresponding to the pI values of the sample ions; hence, on completion of the separation, the ions of the sample, situated in a mobile pH-gradient, are differentiated from one another in sharp zones. The sample may be concentrated by a factor of as much as 1000-10000.

The special fields of application of *isotachophoresis* are the separation of various *peptides*, *metabolites*, *amino-acids* and *nucleotides*.

Theoretical questions of *isotachophoresis*, its instrumentation and the scope of its application are discussed in detail by Everaerts, Beckers and Verheggen .

Analytical Isotachophoresis

Any apparatus that may be employed for gel *electrophoresis* may also be used for analytical *isotachophoresis*. The proteins are first separated in an appropriate buffer without *ampholyte*, and then with the application of an *ampholyte* of the selected pH interval and suitable concentration.

Of the numerous gel compositions that may be employed for *isotachophoresis*, a gel with a 7.5% *acrylamide* content and the following composition has very wide application:

acrylamide solution (T=30%, C=3.33%)	1 volume
gel buffer, pH 8.2	1 volume
sucrose solution (25%)	1 volume
riboflavin solution, 0.004%	1 volume

If it is desired to work with a 5% or a 3.75% gel, 2 or 4 volumes, respectively, of distilled water are added to the composition above.

Compositions of the buffers:

Gel buffer, pH 6.1:

TRIS	5.2 g
1 *M* phosphoric acid	39.0 ml
N, N, N', N'-tetramethylethlyenediamine (TEMED)	0.5 ml
H_2O	to 200.0 ml

Stock solution of leading buffer:

TRIS	5.2 g
1*M* phosphoric acid	39.0 ml
H_2O	to 200.0 ml

Before use it is necessary to dilute the stock solution of the leading buffer, to an extent depending on the composition of the gel:

for 7.5% gel: 1 vol. stock solution + 3 vols. water
for 5.0% gel: 1 vol. stock solution + 5 vols. water
for 3.75% gel: 1 vol. stock solution + 7 vols. water.

Terminal buffer, pH 8.2:

(i) β-alanine	35.8 g
H_2O	to 2000.0 ml
(ii) glycine	30.0 g
H_2O	to 2000.0 ml

The pH of the solutions must be adjusted to 8.2 with TRIS.

The gel solution is poured into 15 × 0.5 cm tubes to a height of 13 cm, gel buffer is layered onto its surface, and *photo-polymerisation* is carried out for I h. After the *polymerisation*, the gel tubes are placed in an apparatus that can be used for rod electrophoresis, and the protein sample, dissolved in terminal buffer, mixed with *saccharose* solution and *ampholyte*, is layered onto the surface of the gel. The upper bath is filled with terminal buffer, and the lower one (including the *anode*) with gel buffer. *Isotachophoresis* is carried out for 2-3 h at constant ionic strength and at 2 mA per gel. After completion, the gels are taken out, fixed and stained by the methods described in connection with IEF.

In isotachophoresis, the length of the protein zone is governed by the concentration of the protein, but is also influenced by the concentration of the ampholyte. A faster or slower separation can be attained, depending on the pH of the leading *electrolyte*. The concentration of the leading and terminal ions, and the distance between them, likewise influence the separation: a low concentration of the leading ions shortens the separation and the protein zones are longer. If the parameters are suitable, *lengthening* the separation does not impair its quality, since the possibility of diffusion is excluded.

Preparative Isotachophoresis

Svendsen and Rose used a preparative column for isotachophoresis. The current is conducted by the *isotachophoretically* migrating ions and the common *counter-ions*. Hence, the ions of the protein sample proceed one after another at the same rate. At a given rate the electric field is inversely proportional to the mobility, and thus, as in the analytical method, identical rates of the zones result in a stepwise increase of the field strength. The ions of the sample therefore cannot pass over into the neighbouring zone because of the lower or higher field.

Various forms of preparative *isotachophoresis* may be selected, and the loading may be about 1000 times that possible in preparative PAGE. In spite of this advantage, *isotachophoresis* has so far been used only to solve research problems, and has not yet gained ground among the routine methods.

7

Peptide Sequence Analysis

Advances in ionisation methods and *instrumentation* that have occurred over the past decade enable MS to play a key role in the emerging area of proteomics. For example, matrix-assisted laser desorption *ionisation* (MALDI) and *electrospray* ionisation (ESI) are now routinely used to ionise peptides and proteins, which can then be aimed at sequencing peptides and/or proteins, determination of post translational modifications, or to study protein–protein complexes. Although various instrument platforms can be used for the mass analysis, over the past decade great strides have been made in the development of TOF instrumentation. Advances in instrument geometry and *electronics* used for TOF MS allow us to interrogate large gas-phase ions with improved resolution and *mass accuracy*. Two extremely important TOF mass spectrometric techniques to the field of *proteomics* are *peptide mass fingerprinting* (PMF)/*protein database* searching and peptide sequencing, and both methods owe their widespread acceptance to the versatility and user friendliness of modern TOF instruments.

PMF, combined with genomic database searching to identify proteins, is now widely used, and PMF based on accurate mass measurement is now a routine TOF analysis. Peptide identification based on PMF/protein database searching is especially powerful when combined with peptide sequencing by tande m MS. For example, determining a partial peptide sequence or identifying a specific post translationally modified a mino acid dramatically increases the confidence level of identification.

For proteins that are not listed in a database, peptide sequencing is essential for protein identification. Peptide sequencing using MS is based on fragmentation of the gas-phase ion. For example, fragmen tation along the peptide *backbone* yields a series of ions separated in mass by an amount equal to the mass of the eliminated residue. Consequently, the ions must be formed with sufficient internal energy to undergo *unimolecular* dissociation (metast able ion decay or PSD), or ions formed with very little initial internal energy (i.e., by ESI) must be activated to energies above the dissociation thres hold. The most commonly used ion-activation meth ods for peptide sequencing are collision-induced dissociation (CID) and surface - induced dissociation (SID); however, photo dissociation has considerable potential for peptide sequencing, especially for peptide sequencing involving TOF instruments.

The physics of CID are not compatible with peptide sequencing via PSD focusing using the standard tandem TOF instrument. For instance, high-energy CID tends to enhance the yield of low *m/z* fragmentions at the expense of higher mass (e.g., more *structurally /* sequence informative) *fragmentions*. Perhaps most troublesome for tandem TOF is the physics of converting kinetic energy into internal energy. CID yields ions having a range of velocities and a broad distribution of arrival times, which limits both *m/z* resolution and mass measurement accuracy of the TOF measurement. Another important consideration for CID of large ions is the effect of center-of-mass collision energy. In virtually all tandem TOF experiments, the mass of the ion undergoing CID is much larger than the target atom. Consequently, only a very small frac tion of the laboratory energy can be partitioned as internal energy of the ion, and as the mass of the analyte ion increases, energy deposition becomes even less efficient.

To circumvent some of the limitations of CID, several groups began investigating SID as an alternative method of ion activation. Under optimised conditions, SID yields very good fragment ion intensities; however, at higher collision energies, SID produces primarily low *m/z* ions, for example, *immonium* ions. As the translational energy of the peptide ion is decreased, surface sticking, or ''*soft-landing*,'' becomes competitive with fragmentation processes. Also, because fragment ions in SID leave the surface with a range of *velocities* and *angles*, careful attention must be paid to instrument geometry.

Several years ago we began evaluating 193-nm photo dissociation as a method to circumvent some of the limitations of CID. Radiation from an ArF excimer (193-nm, 6.43 eV / photon) laser is particularly well suited to *photo dissociation* of peptide ions, because the amide linkage is a natural *chromophore* (λ_{max} ~ 210 nm), and the *photon imparts* sufficient energy to cleave the bond.

Relatively few tandem TOF photodissociation experiments have been reported. This is in part due to logistical issues that must be addressed in a tandem TOF photo dissociation experiment. First, the timing for the *trigger* pulse for the *excimer* laser with respect to the *ionisation/desorption* event is critical. In order to intersect the ion beam with a pulsed laser beam, both the delay generator output and the intrinsic time delay in the laser *trigger* must be stable to ±2 ns. Second, the ions should be *irradiated* in a region where they are spatially separated and well defined, in terms of *m/z*. Third, they must be given sufficient time to undergo *unimolecular* decay in the field-free region before entering the reflectron.

Several years ago we began a series of studies to evaluate photo-dissociation of peptide and protein ions. The instrument consisted of two *collinear* TOF mass *analysers*, with a photo-dissociation region positioned between the two analyzers. The instrument used for the preliminary studies was specifically designed to investigate photodissociation of MALDI-generated peptide and protein ions; however, the instrument did not include a reflectron to mass analyse the resultant *photo fragments*. These studies allowed us to differentiate between fragment ions and neutrals produced by *photodissociation* of the peptide and protein ions. We also showed that *photodepletion* of $[M + H]^+$ ions is a single-photon process, whereas *photodepletion* of fragment neutrals, F_n^0, formed by photoionisation of the $[M + H]^+$ is a two-photon process. Finally, based on the observation that photodepletion occurred within 1 μs of irradiation, it was determined that the rate constant for *photodepletion* was at least 10^6 s^{-1}.

Early photo dissocia tion studies were aimed at small organic molecules, and many of these studies were carried out using Fourier transform mass *spectrometers* (FT-MSs). The FTMS *photodissociation* experiment is relatively simple; a defocused light source is passed through the cell and the photo fragment ions are trapped and detected using conventional methods. The long trapping times (milliseconds to hundreds of seconds) and density of neutral molecules (baseline vacuum of 10^{-6} to 10^{-9} torr) is sufficient for the analyte ion and/or photofr agment ions to undergo ion–molecule reactions, which can complicate the photofragment ion spectrum. Due to the ions being trapped in the FT-MS ion cell and bathed in a continuous radiation field, it is difficult to control the *photoexcitation* process (e.g., single vs. multip hoton exci tation). Multiphoton excitation deposits very large amounts of energy to the *molecule*, resulting in extensive fragmen tation. In addition, *photofragment* ions can also absorb a photon and yield second-generation phot ofragment ions. These complications do not apply to TO FMS photodissociation experiments.

INSTRUMENTATION

The results from our preliminary experiments demonstrated efficient photodissociation of protonated peptide-protein ions and motivated us to design and construct an instrument for mass analysing the photo fragment ions. The instrument used for these studies and it consists of a linear TOF-1 and a *reflectron* that can be used as TOF-2. The effective flight-path length of TOF-1 is 0.5 m and TOF-2 is 3.4 m. Two turbo molecular pumps main tain baseline pressure

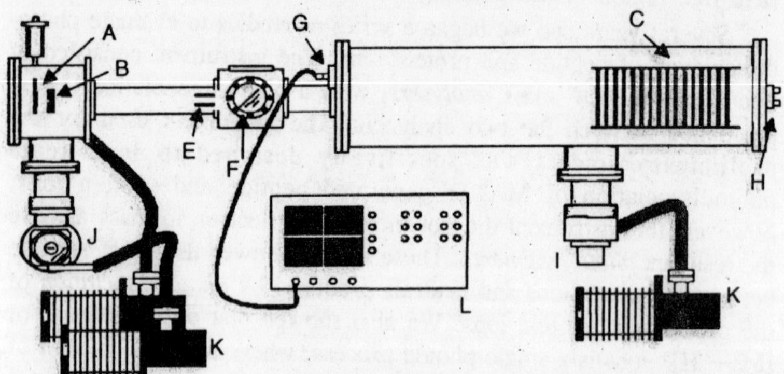

Figure 7.1 : Schematic of the MALDI Photofragment Tandem TOF Instrument (A) Source plate; (B) extraction grids; (C) reflectron; (D) stage manipulator; (E) timed ion selector; (F) quartz window; (G) reflected ion detector; (H) linear ion detector; (I) 170-L/s turbo molecular pump; (J) 330-L/s turbo molecular pump; (K) roughing pumps; (L) 1-GHz digital oscilloscope.

of approximately 2×10^{-8} torr. It is important that such a high vacuum be maintained to eliminate the possibility of ion collisions with background gases, which contribute "*background fragmentation*" to the experiment.

Ions are gen erated in the ion sou rce by MALDI (337 nm) and extracted by pulsing the extraction grid potential to ground. Ions of interest are selected by the *timed-ion* selector and subsequently photodissociated by the *excimer* laser. The output of the *excimer* laser is directed to the mass spectrometer via three *reflective* mirrors (front surface specific to 193 nm), and laser beam size is controlled by a mechanical iris positioned between the final mirror and the *vacuum* window. The laser output energy is controlled by the laser control panel, and fine control is achieved by passing the laser output through a sample cuvette filled with a methanol/water solution. Control of the laser photon density output is important, because at high laser beam energy, the *photofragment* ion spectrum is dominated by low *m/z immonium ions*, and at lower laser beam energy, the abundance of structurally informative backbone and side chain cleavages are increased. Precautions are taken to ensure that the laser beam does not strike the surface of the vacuum chamber because this causes *photoejection* of *electrons* and *neutral* species that can enhance abundances of CID fragment ions.

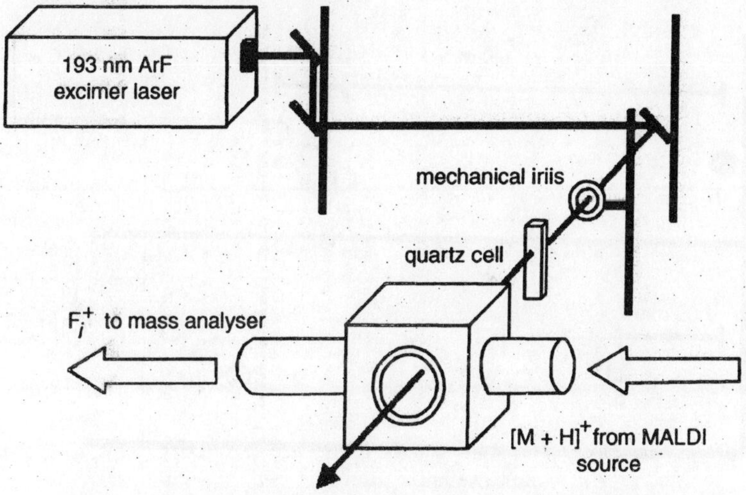

Figure 7.2 : Expanded view of the laser–mass spectrometer interface.

Photofragment ion spectra are calibrated using methods commonly employed for PSD spectra. Briefly, photofragment spectra are taken at a number of PSD *reflectron* ratios and then are truncated to include only that portion of the spectrum within a few *microseconds* of the arrival time of the *parent ion*. The truncated *spectra* are spliced together to create a complete photofragment ion spectrum.

A limitation of PSD experiments is the relatively low mass resolution of the timed-ion selector (MS^1). Currently, the timed-ion selector is limited to a 300-ns pulse. Thus, all ions that arrive at the timed-ion selector within the 300-ns window are allowed to continue toward the *reflectron*, and any *fragment ions* formed by dissociation of these ions are observed in the *photofragment* ion spectrum. Although the mass resolution of the timed-ion selector is low enough that $[M + H]^+$ peptide ions of *m/z* 1000 (20-keV translational energy) and $[M + Na]^+$ ions are not resolved, it is possible to selectively *photoexcite* only the $[M + H]^+$ or $[M + Na]^+$ ion. For example, as the ions approach the laser-ion beam intersection, the $[M + H]^+$ and $[M + Na]^+$ ions are separated by 6.0 mm, and during a single excimer laser pulse (17 ns), the ions travel approximately 1.0 mm. Thus, even if the $[M + Na]^+$ ions are allowed to pass the TIS to the *photodissociation* region, they are not irradiated, and their fragment ions should not appear in the photofragment ion spectrum. It is important that *fragmentation* from ions such as $[M + Na]^+$ be eliminated, because they produce product ions shifted in mass (+23 Da), making spectrum analysis significantly more difficult.

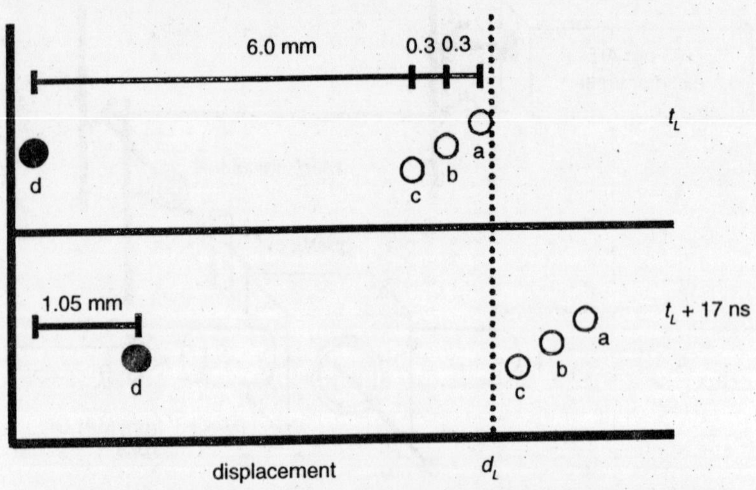

Figure 7.3 : Spatial separation of ions at the laser–mass spectrometer Interface Isotopes of a theoretical peptide $[M + H]^+$ of m/z 1000 (a, b, and c) are separated by approximately 0.3 mm, whereas the $[M + Na]^+$ ion (d) is approximately 6.0 mm behind. The excimer laser is fired at time tL when the $[M + H]^+$ isotopic distribution reaches the point d_L. At time t_L + 17 ns, the excimer laser beam exits the instrument. All ions have advanced approximately 1.05 mm, and only the isotopic distribution of the theoretical peptide $[M + H]^+$ ion, having traversed through point d_L, has absorbed 193-nm photons.

Using the PSD method to analyse photofragment ions has limitations as well. *Metastable ions* that are formed within the first field-free drift region (TOF-1) have the same velocity as their respective parent ions. The metastable ions arising from the ion of interest are activated by the *photodissociation* laser, giving rise to small-neutral loss ions, which complicate the photofragment ion spectra. Our previous work also showed that *metastable* neutral species can be ionised by 193-nm irradiation. The most significant drawback to the *photodissociation* experiment using PSD focusing is the time required for data acquisition and calibration. As with most PSD experiments, the reflectron voltage must be lowered in increments to acquire the entire fragment ion spectrum, so data acquisition must be repeated *n* times, where *n* represents the number of segments needed to stitch the spectrum. To circumvent this limitation, nonlinear reflectron fields have been prop-osed and used with some success. Alternatively, ions could be activated "*in-source*" so that all fragment ions are accelerated to the same final kinetic energy. In this case, acquisition of the photofragment ion spectrum would have the same duty cycle as a conventional TOF mass analysis.

Our most recent photodissociation experiments used an instrument *design* that *negates* the limitations described earlier. A four-grid deceleration acceleration cell (10 cm in length) was added, centered about the *photodissociation* window. The MALDI source potential for these experiments was 15 kV, and 8 kV was applied to the central element of the photodissociation cell, and the outside cell grids are held at ground potential. Ions are separated from their *metastable* decay products as they traverse the decelerating *electric field*, and consequently, only the ions of interest are irradiated at the center of the cell. A fraction of *photoexcited* ions dissociate within 0.5 μs as they traverse the remaining field-free region of the *photodissociation* cell. As ions exit the photodissociation cell, they are then accelerated to near-uniform kinetic energies and *m/z* analysed by TOF-2. Because differences in final kinetic energies between parent and product ions are small, a complete *photofragment* ion spectrum is acquired at a single reflectron voltage. Variation in final *kinetic energies* between photofragment ions is dependent on parent ion mass-to-charge ratio and is accounted for in a new calibration equation:

$$\mathrm{TOF} = a\left(\frac{m_f}{bm_f + c}\right)^{\frac{1}{2}} + t_0$$

Parameters a and c are constant for given mass spectrometer dimensions and source and cell potentials, and parameter *b* is inversely proportional to the *m/z* of the parent ion.

The $[M + H]^+$ ions for a mixture of five known peptides (HLGLAR, des-Arg^9 bradykinin, angiotensin III, bradykinin, and substance P) were sequenced by both PSD focusing and TOF-TOF photodissociation tandem MS methods. All mass spectra and photofragment ion spectra were acquired using sample preparation methods that reduce the abundance of PSD fragment ions (e.g., near-threshold laser power for MALDI) and using fructose sample preparations. CID sp ectra were acquired using an Applied Biosystems 4700 Proteomics Analyzer (Framingham, MA). Instrument conditions used for CID included 1-kV accelerating voltage and medium gas pressure.

MECHANISM OF PHOTODISSOCIATION

Absorption of ultraviolet radiation by a gas-phase ion yields an electronically excited ion, and the photoexcited ion may dissociate directly or decay by radiative and/or nonradiative channels. Although radiative decay of gas-phase ions is an important route for rigid *molecules*, *nonradiative* decay is a much higher probability process

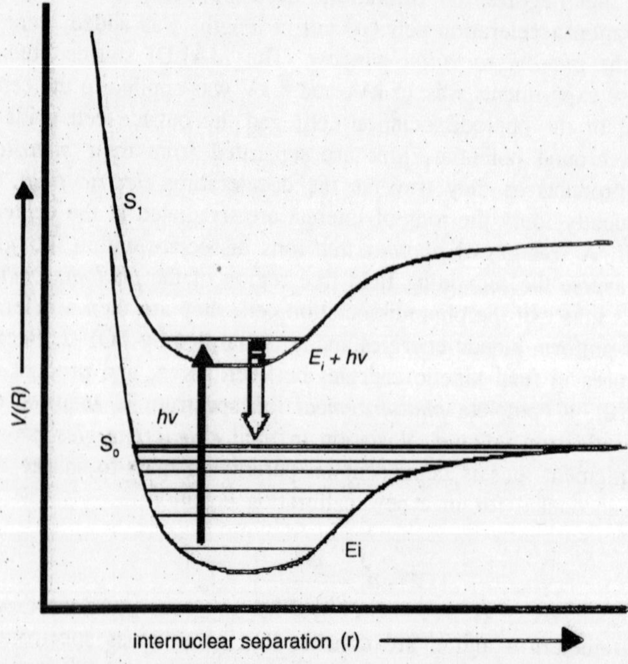

Figure 7.4 : Vibrational predissociation mechanism absorption of a photon of 193 nm is sufficient to induce a Franck Condon transition to the upper electronic state. The ion may then relax via internal conversion to the lower electronic state and fragment.

for peptide ions. Because the gas-phase ion is an isolated system (i.e., energy cannot be transferred to solvent or the local environment), the *photoexcited* ion may partition the electronic energy to vibrational modes, and the *vibrationally* excitedion can then undergo dissociation, provided sufficient energy and time are available. According to RRK theory, the rate for unimolecular dissociation for a vibrationally activated ion having s degrees of freedom and internal energy E is calculated using the following equation:

$$k(E) = V\left(\frac{E - E_0}{E}\right)^{s-1}$$

Rates of dissociation are directly proportional to the amount of excess energy required for bond cleavage ($E - E_0$) and the frequency of the critical oscillator (v). The process of electronic excitation of a specific *oscillator* via single photon absorption may increase the rate of dissociation for fragmentation pathways that are not accessible for lower energy activation methods. *Photofragmention* spectra of peptides activated by 157 nm radiation also provide evidence that *photodissociation* can occur promptly at bonds that are near the chromophore.

PEPTIDE SEQUENCING

For peptide sequencing experiments, it is important to accurately determine the *m/z* values of fragment ions. Most tandem TOF sequencing experiments have average mass errors in the range of 0.1–1.0 Da. Although the mass difference between most of the amino acids is more than 1 Da, there are several cases in which mass differences are much less. Leu and Ile have identical masses (113.084 Da) and cannot be distinguished based solely on fragmentation along the peptide backbone. Gln and Lys differ by only 0.036 Da (128.059 and 128.095 Da, respectively), so a mass measurement accuracy of ±0.01 is required to distinguish these residues.

The list fragment ion mass measurements for the peptide HLGLAR acquired during a PSD focusing photodissociation experiment. The average mass measurement error of 0.028 Da is sufficient to elucidate the composition of this peptide (with the exception of differentiating between Leu and Ile) from either terminus. Both mass accuracy and resolution are higher for photodissociation TOF-TOF experiments.

The utility of photodissociation tandem TOF for peptide sequencing is illustrated by the photofragment ion spectra of bradykinin contains

Table 7.1 : Fragment Ion Mass Accuracy Data for the Hexapeptide Hlglar Acquired on the MALDI Photofragment TOF Instrument.

TOF (ns)	PSD ratio	Ion identity	Meas. *m/z*	Calc. *m/z*	Error
49235.53	1.000	$[M + H]^+$	666.405	666.405	0.000
47248.32	0.870	y_5	529.395	529.346	-0.049
49195.00	0.770	y_5-NH_3	512.286	510.304	0.034
49110.40	0.770	$b_5 + H_2O$	510.387	492.293	-0.083
49192.09	0.740	b_5	492.275	492.293	0.018
47897.94	0.740	a_5	464.347	464.299	-0.048
47016.22	0.700	b_4	421.262	421 .256	-0.006
48858.32	0.635	y_4	416.284	416.262	-0.022
47941.37	0.635	y_4-NH_3	399.302	399.236	-0.066
47616.95	0.635	a_4	393.294	393.261	-0.033
48375.26	0.560	y_3	359.259	359.241	-0.018
48517.89	0.530	y_3-NH_3	342.232	342.214	-0.018
47509.16	0.500	b_3	308.163	308.172	0.009
47372.89	0.410	b_2	251.110	251.151	0.041
47149.81	0.378	y_2-NH_3	229.071	229.130	0.059
48868.05	0.340	a_2	223.116	223.156	0.040
46084.94	0.240	b_1	138.073	138.067	-0.006
47695.02	0.180	R	112.082	112.087	0.005
48541.49	0.170	A_1(H)	110.074	110.072	-0.002
49028.12	0.130	L	86.088	86.097	0.009
				Average	0.028
				Std. Dev.	0.024

the PSD focusing photofragment ion spectrum of bradykinin (RPPGF-SPFR) taken from a five-peptide mixture. As in the previous example, *immonium ions* corresponding to proline, phenylalanine, and *arginine* are observed, and a complete set of *a*-type ions (except a_7) are observed. In addition, several b-type ions are observed as well, notably the $b_8 + H_2O$ ion at *m/z* 904. Ions corresponding to charge retention by the *C-terminus* are prominent in this spectrum, as evidenced by several prominent *y*-type ions. Because there are arginine residues at either terminus of this peptide, it is reasonable to expect significant populations of fragment ions exhibiting C-terminal charge retention from this peptide. In addition to simple backbone cleavage products,

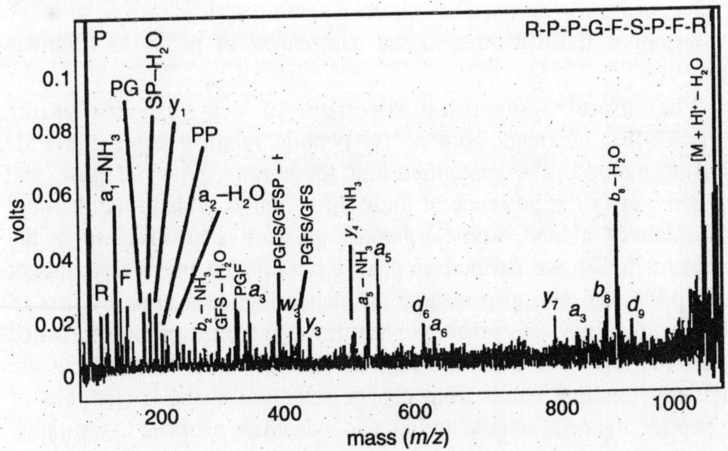

Figure 7.5 : Post-source decay (PSD) focusing photofragment ion spectrum of bradykinin (RPPGFSPFR) taken from the five peptide mixture.

numerous side-chain cleavage products are observed in this spectrum, notably d_6, d_9, and w_3. We also observe internal fragmentation in this spectrum, which is typical for proline-containing peptides. Internal fragmentation is highly useful in reconstructing peptide sequence when a complete series of backbone cleavage products is not observed. For instance, using just the a-type ions observed in this photofragment ion spectrum, the sequence would be assigned "RPPGFSXXR," where XX can be either FP or PF, but knowledge of one of the internal cleavage ions such as FSP, GFSP, or PGFSP indicates the sequence must be RPPGFSPFR. Alternatively, this information could be extracted from C-terminal sequence information given by the y-type ion series. In this case, we are able to assign the sequence as XPXXXS-PFR from the observed *y*-ions. Thus, in the case of bradykinin, the information obtained from the *a*- and *y*-type fragment ions is sufficient to *unambiguously* assign the complete sequence.

Note that contributions from multiple fragmentations, such as internal cleavage ions and small neutral losses, are in much lower abundance in this spectrum. Products of prompt *photodissociation*, such as *a*-type sequence informative ions and *d*-type side-chain cleavage ions, are now the major contributors to the photofragment ion spectrum. The CID spectrum of this species is shown in Fig ure elsewhere in this chapter. Note the appearance of several ion types having charge retention on either side of the *dissociating* bond and internal fragment ions and other unidentified species. Although much information is contained within this spectrum, its utility for de novo sequencing is low because of difficulties in the *assignment* of peaks to fragment ions.

The photofragment ion spectrum of des-Arg[9] bradykinin (RPPGFSPF) obtained from a five-peptide mixture using the PSD focusing method. The *immonium* ions for *proline*, *phenylalanine*, and *arginine* verify the presence of these three amino acids in the peptide. A number of a- and b-type fragment ions are also observed in the spectrum, which we attribute to charge retention at the N-terminus of the peptide ion. We also observe sequence ions that arise by loss of NH3 from either the primary photofragment ions or the activated metastable ions, and these are labeled Fi-NH3 ions. Also present are fragment ions that result from charge retention at the C-terminus of the peptide ion or at a basic amino acid side chain near the C-terminus. For example, y_1, y_2, and y_3 are observed. In addition to simple backbone cleavage reactions (e.g., *a-*, *b-*, and *y*-type ions), we also observe side-chain *cleavage* ions. For instance, v_4 and w_6 arise by side-chain cleavage products in which the charge is retained by the C-terminus. In addition to the *v*– and *w*-side-chain cleavage ions, side-chain cleavage product ions coresponding to N-terminal charge retention, d_2, d_3, and d_6 are also observed.

The TOF-TOF photofragment ion spectrum of des-Arg[9] bradykinin is shown in figure elsewhere in this chapter. Photofragment ions arising from multiple *dissociations* and/or fragmentation of *metastable* ions are in much lower abundance. C-terminal charge retention products such as *phenylalanine immonium* ion and y-type ions are observed in lower abundance also. Analysis of these promptly formed photofragment ions provides evidence that N-terminal charge retention is dominant in the initial population of MALDI formed ions. Through comparison with results from the PSD focusing method, we postulate that b- and y-type ions are more likely to be formed from lower energy, longer-lived dissociating ions, as they appear in much lower relative abundance when the time allotted for observation of photofragment ions is decreased from 10 /is to 0.5 /is.

The CID spectrum of fibrinopeptide A (ADSGEGDFLA-EGGGVR). The spectrum consists of mostly y-type sequence ions and w-type side chain cleavage ions. Note the dominance of the y-type ions corresponding to backbone cleavage at acidic residues. These same *dominant* ions are observed in the PSD focusing photofragment ion spectrum.

When photoactivation is used, side- chain cleav age product w_{8a}

become s more prominent in the tandem mass spectrum. The TOF - TOF photo fragment ion spectrum is shown in figure . Strong signa ls from *y*- type ions are no longer observed, rather *x*-type sequence ions are the most comm on. Side -chain cleavage products in this spectrum are observed in abundance similar to that of the other sequence informative ions. Comparison of these spectra further confirm s that y- type ions are formed subsequent to internal energy reorganisation, requiring a greater fragmentation time scale. Through comparison of the PSD focusing and TOF - TOF methods, we observe that photo activation using 193 - nm radiation causes both prompt fragmentations at the sites of the chromophores, as well as *rotationally* and *vibrationally* excited species, which dissociate over longer time scales.

Although the isomass residues *leucine* and *isoleucine* cannot be distinguished on the basis of m /z, these residues can be distinguished by side - chain cleavages. For example, *leucine* exhibits a loss of 42 Da and *isoleucine* loses 28 Side - chain cleavage products are rarely observed by low-energy activation or in PSD spectra of MALDI-formed ions; however, side-chain cleavage reactions are frequently observed by high-energy CID and *photodissociation* The TOF-TOF *photofragment* ion spectrum of peptide HLGLAR is shown in Figure. Bccausc thc most basic residuc in this pcptidc is argininc, primarily C-terminal charge-retaining photofragment ions are observed. Side-chain cleavage ions w_{3a} and w_{5a} appear in similar abundance to other sequence informative ions, which can be used to *unambiguously* assign *leucine* to the second and fourth amino acid residues. A complete x-type ion series is observed, as well as other sequence informative ions. The TOF-TOF *photofragment* ion spectra of HLGLAR, *fibrinopeptide* A, and des-arg^9-*bradykinin* serve to further illustrate the analytical utility of de novo peptide sequencing using 193-nm *photoactivation*. Peptides that are most basic at the N-terminus produce predominantly *a*-type ions. C-terminal charge-carrying peptides, such as peptides that result from protein digestion with trypsin, *photodissociate* into x-type ions. Interpretation of tandem MS data is simplified when a single complete sequence ion series dominates the mass spectrum. Further sequence information is obtained in the form of side-chain cleavage ions for both N- and C-terminal charge carriers.

To further evaluate the *photodissociation* tandem TOF for peptide sequencing, we analysed a tryptic digest of the protein cytochrome C. A-1 zl aliquot of the tryptic digest (0.5 mg/ml) was mixed with matrix and spotted on the MALDI sample stage. It should be stressed that no sample pretreatment was performed; all buffers, salts, and all other impurities present in the digest solution are present in the MALDI sample spot. The MALDI mass spectrum of this sample. Two peptides were selected, those at *m/z* 1633 and 1168. We previously assigned the sequence of the peptide at *m/z* 1633 to a heme-containing peptide (CAQCHTVEK þ heme) based on accurate mass measurements of the parent ion, and this assignment is consistent with the *photofragment* ion spectrum. Note that the dominant *photofragment* ion is the heme at *m/z* 616. Both *a*- and *b*- type cleavages are observed in the spectrum, but there are too few fragment ions to completely sequence this peptide. However, confirmation of the presence of the heme group supports our earlier assignment.

The PSD focusing *photofragment* ion spectrum of the *m/z* 1168 peptide ion (TGPNLHGLFGR) of cytochrome C. This spectrum contains signals that correspond to the immonium ions of *asparagine* and *phenylalanine*. On the basis of *a*-, *b*-, and *y*-type fragment ions, the sequence of this peptide is assigned as TGPNXHGXFGR, where X is either *leucine* or *isoleucine*. This spectrum also contains a number of internal and side-chain cleavages. Of particular interest are the d8 and w_{7a} ions, both of which correspond to leucine residues. Thus, we can assign positions 5 and 8 to leucine and the sequence can be unambiguously assigned to TGPNLHGLFGR.

The TOF-TOF photofragmention spectrum of *phosphorylated angiotensin* II (DRVpYIHPF) is displayed in Figure elsewhere in this chapter. Complete *amino acid* sequence information is obtained from *a*- and *y*-type ion series. The *isoleucine* at the fifth position is *unambiguously* identified using the d_{5a} ion. The identity of the modified amino acid is revealed by the *phosphorylated tyrosine immonium* ion, and its position is clearly denoted by the a-type ion series. Photofragment ion spectra of *phosphorylated* peptides acquired using the PSD focusing method include ions identified as peptide fragments with loss of phosphoric acid. It was uncertain as to whether these fragment ions arise from photodissociation of the phosphorylated peptide ion or its phosphate-loss *metastable* ion. In the TOF-TOF *photodissociation* experiments, sequence ions with *phosphate* loss are not observed since the *phosphate*-loss metastable ions are not activated.

Photodissociation of MALDI-generated peptide [M þ H]þ ions with 193-nm excimer radiation results in significant fragmentation. Current results demonstrate that this fragmentation may be used to reconstruct peptide sequences. In addition, *photodissociation* results in significant numbers of high-energy side-chain cleavages, which are particularly useful for differentiating between isomass residues. Our results demonstrate the ability to select peptides of interest from a digest solution and sequence them in the mass *spectrometer* with no prior sample treatment.

Also, photodissociation is a reliable method for probing *posttranslational* modification.

Within the last decade, MS has become the method of choice for protein identification. Whole cell lysates, components of protein complexes, and proteins isolated by *immunoprecipitation* or affinity chromatography are now being routinely studied using this technique. The key element in this success derives from the ease of peptide sequence determination based on interpretation of peptide fragmentation spectra. Similarly, these techniques are also used for recombinant protein characterisation and for structure elucidation of *biologically* active small peptides, such as *toxins*, *hormones*, and *antibiotics*. Peptide sequence and structural data can be obtained by collisional activation of selected singly or multiply charged precursor ions. The principles of high-energy collisional activation have been established in combination with fast atom *bombardment* (FAB), liquid secondary ion MS (LSIMS) *ionisation* on four-sector instruments, also with MALDI on sector-orthogonal-acceleration-time-of-flight (TOF) hybrid *tandem* instruments, and on a TOF-TOF *tandem* mass *spectrometer*. Low-energy CID spectra are acquired on all the other *hybrid tandem* instruments, for example, *quadrupole-orthogonal*-acceleration-TOF (QqTOF) instruments, by triple-*quadrupole* mass spectrometers and ion traps regardless of the ionisation method applied. The unimolecular decomposition of peptide ions generated by MALDI may be detected and recorded by PSD analysis in MALDI-TOF instruments equipped with reflectron. Besides traditional CID, Fourier transform MS (FT-MS) instruments also used sustained off-resonance irradiation (SORI) and infrared multiphoton dissociation (IRMPD) to generate MS/MS spectra of larger multiply charged molecules, and electron-capture dissociation (ECD) has been introduced for structural characterisation of multiply charged ions. A novel version of this MS/MS method is hot ECD (HECD), when multiply charged *polypeptides* fragment upon capturing approximately 11-eV electrons. Another novel MS-MS technique, electron-transfer dissociation (ETD) can be performed in radiofrequency (RF) *quadrupole* linear ion traps. In ETD, singly charged anions transfer an electron to multiply charged peptide ions and induce fragmentation analogous to that observed in ECD. In general, in all kinds of MS/MS experiments, the mass *spectrometric* detection sensitivity is high, and sequence information is obtainable in peptide mixtures, even for peptides containing unusual or covalently modified residues. Peptides yield a wide array of product ions depending on the quantity of vibrational energy they possess and the time window allowed for *dissociation*. The ion types formed and the abundance pattern observed are influenced by the peptide sequence, the ionisation technique, the charge state, the collisional energy (if any), the method of activation, and the type of the analyser. In this chapter, a comprehensive list of common peptide product ions is presented, and the fragmentation differences observed in different MS/MS experiments are discussed in a qualitative manner. Examples are shown on how the MS/MS data are used for de novo sequence determination, protein identification, and assignment of covalent modification sites.

PEPTIDE FRAGMENTATION PROCESSES

Peptides produce fragments that directly provide information on their amino acid composition: immonium (and related) ions and ions that are the products of side-chain losses from the precursor ion.

Amino acids may form *immonium* ions with a structure of $^{+}NH_2 = CH\text{-}R$, where the mass and the stability of the ion depend on the side-chain structure. Immonium ions sometimes undergo sequential fragmentation reactions yielding ion series characteristic for a particular *amino acid*. In high-energy CID experiments, these fragments are usually very reliable and offer a wealth of information. PSD spectra also feature relatively abundant *immonium* and related ions, especially when the decomposition is enhanced by collisional activation. Low-energy CID experiments on singly or multiply charged *precursors* generated by FAB or *electrospray* ionisation (ESI) usually yield some information on the amino acid composition of the peptide, whereas MALDI low-energy CID spectra acquired on QqTOF instruments (or FT-ion cyclotron resonance [ICR]) feature very few and very weak *immonium* ions. This compositional information is usually completely lost when the CID experiment is carried out using an ion trap, in which fragments below approximately one-third of the precursor ion mass are not detected. *Immonium* ions are not formed at all in ECD experiments.

Table 7.2 : Immonium and Related Ions Characteristic of the 20 Standard Amino Acids

Amino acid	Immonium and related ion(s) masses		Comments
Ala	44		
Arg	129	59, 70, 73, 87, 100, 112	129, 73 usually weak
Asn	87	70	87 often weak, 70 weak
Asp	88		Usually weak
Cys	76		Usually weak
Gly	30		
Gln	101	84, 129	129 weak
Glu	102		Often weak if C-terminal
His	110	82, 121, 123, 138, 166	110 very strong 82, 121,123,138 weak
Ile/Leu	86		
Lys	101	84, 112, 129	101 can be weak
Met	104	61	104 often weak
Phe	120	91	120 strong, 91 weak
Pro	70		Strong
Ser	60		
Thr	74		
Trp	159	130, 170, 171	Strong
Tyr	136	91, 107	136 strong, 107, 91 weak
Val	72		Fairly strong

The other ion type that provides information on the *amino acid* composition of the peptide is formed from the precursor ion via side-chain loss. This dissociation process is very characteristic in high-energy CID spectra obtained on four-sector mass spectrometers with FAB or LSIMS ionisation but plays a far less significant role in other CID experiments, even in MALDI high-energy CID spectra. However, some residues usually undergo this kind of fragmentation, so their presence will be indicated. Met-containing peptides may feature a loss of 47 Da (CH_3S) from the precursor ion. When this residue is oxidised, even the sequence ions containing the *methionine-sulfoxide* undergo an extensive fragmentation via *neutral* loss of 64 Da (CH_3SOH) in most MS/MS experiments. Upon MALDI ionisation, due to the cleavage of the α, β-bond, Phe and Tyr may lose 91 and 107 Da, respectively. Side-chain losses and side-chain fragmentation also have been reported in ECD experiments. Basic amino acids produce the most significant fragmentation: His-containing peptides display an 82 Da loss (—$C_4H_6N_2$), whereas Arg-containing molecules may lose 101, 59, 44, and 17 Da correspo nding to $C_4H_{11}N_3$, CH_5N_3, CH_4N_2, and NH 3, respec tively. Other residues exhibiting such fragmentation in ECD experiments were Asn and Gln: — 45 Da (CH_3NO), Lys: — 73 Da ($C_4H_{11}N$), Met: —74 (C_3H_6S), Trp: — 130 Da (C_9H_8N), and — 116 Da (C_8H_6N), Phe: — 92 Da (C_7H_8), and — 77 Da (C_6H_5); and Val /Leu: — 42 Da (C_3H_6), and — 43 Da (C_3H_7).

Table 7.3 : Rules for the Calculation of Fragment Ion Masses.

	Mass calculation	
Fragment	***Using residue weights***	***From other fragments***
a_i	Σresidue weights —27	b_i - 28
b_i	Σresidue weights +1	MH^+ + 1 - y_{n-i}
c_i	Σresidue weights +18	b_i + 17
d_i	Σresidue weights —12 — side chain	a_i - (R_i - 15)
for Ile	Σresidue weights —55 or —41	a_i - 28 or -14
for Thr	Σresidue weights —43 or —41	a_i - 16 or -14
for Val	Σresidue weights —41	a_i - 14
v_i	Σresidue weights —side chain +18	x_{i-1} + 29
w_i	Σresidue weights —side chain +17	x_{i-1} + 28
for Ile	Σresidue weights —side chain +31 or +45	x_{i-1} + 42 or +56
for Thr	Σresidue weights —side chain +31 or +33	x_{i-1} + 42 or +44
for Val	Σresidue weights —side chain +31	x_{i-1} + 42
x_i	Σresidue weights +45	y_i + 26
y_i	Σresidue weights +19	MH^+ + 1 - b_{n-i}
Y_i	Σresidue weights +17	y_i-2
z_i	Σresidue weights +2	y_i-17
Internal fragments		
b-type	Σresidue weights +1	
a-type	Σresidue weights -27	

While the ions discussed above provide compo sition information, all the other signa ls in MS /MS *spectra* provide information on the amino acid sequence. All primary cleavages occur along the *peptide* backbone. Most frequently the dissociation reaction occurs at the peptide bonds. When the *proton* (charge) is retained on the N-terminal fragment, **b** ions are formed with the structure: H_2N-CHR_1-CO-. . .-NH-$CR_i$$CO^+$ (rules for the calculation of major fragment ion masses are presented in Table elsewhere in this chapter).

These sequence ions (and all the other N-terminal fragments) are numbered from the N-terminus. Normally, fragment $\mathbf{b}_2$ is the first member of this series. However, the presence of N-terminal modifications, for example, *acetylation* leads to the formation of stable $\mathbf{b}_1$ ions. If the *proton* (charge) is retained on the *C-terminal moiety* with H-transfer to that fragment, a **y** sequence ion is formed with the structure $^+NH_3$-CHR_{n-i}-CO-. ..-NH-CHR_n-COOH. This ion series (and all other *C-terminal* ions) is numbered from the *C-terminus*. High-energy CID spectra may also display Y fragments formed by H transfer away from the *C-terminal* fragment, with the structure {NH=CR_{n-i}-CO-...-NH-CHR_n-COOH}H^+. Pro residues often produce abundant Y ions (i.e., they feature a doublet separated by 2 Da). Alternative sequence ion series a and x are formed when cleavage occurs between the *ci-carbon* and the *carbonyl*-group, with structures H2N-CHR1-CO-...-$^+$NH / CHRi and {CO=N-CHRni-CO-;...-NH-CHRn-COOH}H^+, respectively. Alternatively, when the fragmentation occurs between the ci-carbon and the amino group **c** and **z** or **z** + 1 ions are generated, with structures H_2N-CHR_1 -CO-...-NH-CHR_iCO-NH_3^+, {HC(=$CR'_{n-i}$$R''_{n-i}$)-CO-...-NH-$CHR_n$-COOH}$H^+$, and {$CHR_{n-i}$-CO-. . .-NH-$CHR_n$-COOH}$H^+$, respectively. Obviously, the only imino acid, Pro, cannot undergo this type of bond cleavage, so this residue will not yield **z** fragments and amino acids preceding Pro residues will not form **c** ions.

The product ions that are observed in any given spectrum are usually controlled by the basic groups present, such as the amino terminus itself, the "-amino group of Lys, the *imidazole-ring* of His, and the *guanidine* side chain of Arg. This is because these groups are protonated preferentially (i.e., the charge is localised on the basic residues) and fragments containing them retain the charge and tend to dominate the spectrum. For example, tryptic *peptides* usually exhibit abundant *C-terminal* sequence ion series. However, when a *tryptic peptide* with C-terminal Lys contains a His residue close to its N-terminus, the ion series observed may be controlled by the latter, and in such a case, the spectrum will be dominated by N-terminal sequence ions. In general, Arg overcomes the influence of other amino acids, while His and Lys represent similar basicity in the gas phase.

Certain other amino acids may also promote fragmentation reactions. For example, the presence of Pro in a sequence facilitates cleavage of the peptide bond N-terminal to this residue, because of the slightly higher *basicity* of the imide nitrogen, yielding very abundant y fragments. Similarly, cleavage at the C-terminus of Asp residues is favored because of protonation of the peptide bond by the amino acid side chain. This latter effect is especially profound in MALDI-CID and PSD experiments in which abundant **y** fragments are generated via Asp-Xxx bond cleavages. It has been reported that in MALDI low-energy CID of Arg-containing *peptides* when using laser energies below a *threshold* activation level, these bonds will be cleaved exclusively.

In addition, the types of ions observed in MS/MS experiments strongly depend on how the *unimolecular* dissociation processes are induced. In general, fragments **a**, **b**, and **y** are observed in all kinds of CID experiments and in PSD spectra. Additional backbone cleavage product ions are mostly detected only in high-energy CID experiments. Low-energy CID and PSD *spectra* almost never feature **x** and **z** + 1 ions, and some data suggest that the ions at *m/z* **y** - 17 observed in these experiments are the result of ammonia loss from *amino* acid side-chains (Arg, Lys, Asn, or Gln) rather than from the newly formed N-terminus. Occasionally, low-energy CID and PSD spectra may feature c ions, mostly when the charge is preferentially retained at the N-terminus. Interestingly, it has been reported that low-energy CID spectra acquired on a QSTAR (QqTOF) instrument always feature an abundant $\mathbf{c}_1$ fragment if the second amino acid is a *glutamine* residue.

Electron capture leads to entirely different dissociation chemistry. Thus, ECD and ETD spectra display almost exclusively c and z + 1(z·) ions, although some **a** + 1 (**a**·) and **y** fragments may also be detected. For reasons discussed earlier in this chapter, there is usually no cleavage at the N-terminus of Pro residues in ECD and ETD experiments.

The "*major*" sequence ions, **a**, **b**, and **y** may undergo further dissociation reactions, usually via the loss of small neutral molecules. As mentioned already, satellite ions at 17 Da lower mass are produced via ammonia loss from Arg, Lys, Asn, or Gln side chains or at a much lower extent via cleavage of the N-terminal amino group. A loss of 18 Da indicates elimination of a water molecule from the structure. Hydroxy amino acids Ser and Thr, as well as acidic residues Asp and Glu, normally undergo this type of reaction. However, it has been reported that in ion traps, peptides lacking these residues may lose water via a rearrangement reaction. Arg-containing fragments may produce *satellite ions* at 42 Da lower mass, most likely corresponding to the loss of NH=C=NH as a neutral moiety. In addition, any fragment that contains *methionine sulfoxide* may feature abundant satellite ions at 64 Da lower mass, as mentioned earlier. Peptide fragments containing more than one residue capable of undergoing such dissociation reactions frequently yield series of satellite ions because of the various combinations of multiple neutral losses. In addition, in some cases, especially with multiple Arg residues, the relative intensity of sequence ions may diminish or they may completely "*disappear*" while satellite ions of high abundance are detected. These satellite ions can be observed in all kinds of MS/MS experiments, except in ECD and ETD.

Some satellite fragments are characteristic to high-energy CID, although w ions (definitions below) also have been observed in HECD experiments and in conventional ECD spectra of some peptides. HECD of renin substrate also yielded two **d** ions (definition below), so far a unique observation. Fragments d: $\{H_2N\text{-}CHR_1\text{-}CO\text{-}\ldots\text{-}NH\text{-}CH{=}CHR'_i\}H^+$, and **w**: $\{CH({=}CHR'_{n-i})\text{-}CO\text{-}\ldots\text{-}NH\text{-}CHR_n\text{-}COOH\}H^+$ are formed when the fragmentation occurs between the β- and γ-carbons of the side chain of the C-terminal amino acid of an *a* + 1 ion or the N-terminal amino acid of a *z* + 1 fragment, respectively. These satellite fragments permit the *unambiguous* differentiation of isomeric amino acids Leu and Ile. The **d** and **w** ions (!) of Ile are 14 and 28 Da higher than those of the Leu residue, depending on whether the methyl or the ethyl group is retained on the β-carbon, with the lower mass product ion being dominant. *Aromatic amino acids* usually do not produce these fragment ions because of the strong bond between the *aromatic ring* and the *β-carbon*, but sometimes the cleavage may occur in the side chain of the adjacent amino acid. It has been reported that w-type product ions may form this way. Obviously, Pro, the only imino acid, also cannot undergo this cleavage, but it usually yields an abundant w ion that is formed via a different mechanism. High-energy CID experiments may also yield another set of C-terminal ions, the v-fragments: $^{+}NH_2 = CH\text{-}CO\text{-}NH\text{-}CHR_{n-1}\text{-}CO\text{-}NH\text{-}CHR_n\text{-}$

COOH. Pro cannot yield this production. In general, the presence of a basic residue in the fragment (i.e., preferential charge retention at the C-terminus) is essential for the formation of **v** and **w** ions. Similarly, the formation and further dissociation of **a** + 1 ions requires preferential charge retention at the N-terminus. This can be accomplished by the presence of a basic amino acid, or sometimes the basicity of the N-terminus itself is sufficient for **d** ion production. The formation of another N-terminal satellite ion, the **b** + H_2O fragment, is also dependent on the presence of a basic amino acid at the N-terminus. These ions are formed via a rearrangement reaction; "*peeling off*" the C-terminal amino acids one by one. Usually, a one or two amino acid "*loss*" can be detected. These fragments are typical of all CID experiments and PSD spectra. ECD-generated fragments do not feature most of the satellite ions discussed already. However, they may display side-chain losses and "losses of some low-molecular-weight species such as H_2O, CH_3., C_3H_6, $CONH_2$".

Peptides may undergo dissociation processes in which both termini are removed and two or more internal residues form a b-type ion. These species may also yield *satellite ions* losing carbon monoxide, ammonia, or water, just like the *sequence* ions. These product ions are called internal ions and are labeled with the one-letter code of the *amino acids* included. *Abundant* internal ions can be observed in all MS/MS experiments, with the exception of ECD and ETD spectra, which usually lack these ions completely. Pro residues tend to promote this type of dissociation process; complete series of internal fragments are frequently observed in CID and PSD spectra featuring Pro at their N-termini. Acidic residues Asp and Glu also appear to stabilise internal fragments.

In general, high-energy CID offers the most comprehensive information on peptide sequences. It features all the ions listed earlier in this chapter and gives reliable composition information in form of immonium ions and side-chain losses, as well as sequence and satellite ions that permit the differentiation of *isomeric* Ile and Leu *residues*. This wealth of information makes high-energy CID spectra much more complex than their low-energy counterparts or PSD data. Such detailed information may not be necessary when the purpose of the study is protein identification by database searching; and MS/MS methods that yield fewer fragments and, thus, divide the limited signal less could be advantageous for low-level sample analysis. However, high-energy CID data may prove essential for de novo sequence determination and structure *elucidation*, as well as site assignment of covalent modifications. ECD may provide complete sequence coverage for large peptides, as well as differentiate between isomeric Ile and Leu. Importantly, ECD's and ETD's unique fragmentation process also permits the *localisation* of labile *posttranslational* modifications.

De Novo Sequence Determination

A PSD spectrum of a tryptic peptide was selected as the first example to illustrate the sequencing process. These data show that even this most *fastidious* method, with relatively low resolution and mass measurement accuracy for the fragment ions, may yield sufficient information for de novo sequence determination. The interpretation process discussed here also applies to low- and high-energy CID spectra. Obviously, in CID experiments, the fragmentation will be somewhat different, as discussed already.

Furthermore, both the resolution and the accuracy of the mass measurement should be significantly better in spectra obtained with tandem instruments such as the QqTOF hybrid instrument or the MALDI-TOF-TOF tandem mass spectrometer.

Figure elsewhere in this chapter shows the PSD spectrum of a tryptic peptide with MH^+ at *m/z* 1187.6 (1188.3 MH^+ average), isolated from an in-gel digest of the 50-kDa subunit of the DNA polymerase a of Schizosaccharomyces pombe. The immonium ions unambiguously identified the presence of Arg (*m/z* 70, 87, 112), Ile/Leu (*m/z* 86), His (*m/z* 110 and 166), Phe (*m/z* 120), and Tyr (*m/z* 136). These residue weights added together plus the water and the additional proton account for 735 Da, and the "*leftover*" 452 Da could correspond to 65 theoretically possible amino acid compositions. Thus, determining the amino acid composition was not possible, but identifying the terminal amino acids was straightforward. This was a tryptic peptide, featuring a $\mathbf{y}_1$ ion at *m/z* 175, revealing a C-terminal Arg, and the corresponding b_{n-1} ion was also detected at *m/z* 1014. Complementary **b** and **y** ions are related, and their masses can be calculated using the $\mathbf{y}_i + \mathbf{b}_{n-i}$ / $MH^+ + 1$ formula. There was a very abundant ion at *m/z* 1073 that corresponded to a 115-Da loss from the molecular ion. Considering the above discussed correlation, this mass difference is consistent with a $\mathbf{b}_1$ mass of 116 Da, which identifies the N-terminal residue as Asp. It has been observed that cleavage at the C-terminus of Asp residues is preferred during PSD fragmentation processes, yielding abundant y-ions. The next step is to find other sequence ions, perhaps by looking for ions that exhibit neutral losses characteristic of particular ion types. Carbonyl losses could identify b-type ions, both sequence or internal fragments. This spectrum shows some abundant ions that are good b-type candidates, such as *m/z* 253, 285, 398, 416, 535, 640, and 787. The last two are 147 Da apart, corresponding to a Phe residue, and the **y** ion complementary to the 640 b ion occurs at *m/z* 548. The ion at 535 obviously does not fit in the series as there is no amino acid with a 105-Da residue weight. However, there is an abundant ion at *m/z* 503, displaying water loss that is also characteristic of b ions (when containing *hydroxy* or *acidic amino acids*). The mass difference between this ion and the **b** ion at *m/z* 640 is 137, corresponding to a His, and the complementary y fragment is indeed detected at *m/z* 685, along with an ion at *m/z* 668 due to ammonia loss from the y ion. By continuing to "*peel off*" the amino acids, the next **b**-ion is at *m/z* 416 (i.e., 87 Da apart), indicating a Ser residue and explaining the water loss from the **b** ion. In addition, the corresponding y-ion is detected at *m/z* 772. The ion at *m/z* 398, the next candidate, does not fit in the series but may be formed via water loss from *m/z* 416. The ion at *m/z* 285 is 131 Da lower than the already identified **b** ion at *m/z* 416, which may indicate a Met residue. However, there is no other ion (i.e., *immonium ion*, *side-chain loss*, **y** ion), supporting this conclusion; thus, this ion was classified as an internal fragment. The adjacent residue to the Ser is a Tyr as revealed by the mass difference between the b ions at *m/z* 416 and 253. Because the N-terminal residue has already been established as Asp, and 253–116($\mathbf{b}_1$) = 137, the second amino acid must be a His residue. Thus, the N-terminal sequence was determined as Asp-His-Tyr-Ser-His-Phe. Considering these amino acids, plus the C-terminal Arg, the gap that has to be filled is 226 Da. Because it is known from the *immonium ions* that the sequence contains a Leu/Ile with a residue weight of 113, the other missing amino acid must also be either a Leu or an Ile. Indeed, there is a y_2 ion at *m/z* 288, a y_2-NH_3 ion at *m/z* 271, and the corresponding b ion at *m/z* 900. Thus, the sequence of this peptide is AspHis-Tyr-Ser-His-Phe-Leu-Leu-Arg. Note that PSD fragmentation does not allow differentiation between the isomeric Leu/Ile residues. The b-type ions listed above that were not identified as sequence ions, such as *m/z* 285, 398, and 535, are internal fragments corresponding to HF, HFL, and YSHF sequences, respectively. Similarly ions at *m/z* 301 (HF), 372 (SHF), and 672 (HYSHF) are b-type internal fragments; only the a-type fragments formed by the carbonyl loss were not detected. The presence of basic His residues "justifies" why more abundant **b** and internal ions than C-terminal fragments were observed.

The second example is a low-energy CID spectrum acquired by LC/MS/MS analysis on a QqTOF (QSTAR) mass spectrometer in an information-dependent fashion (i.e., the *precursor* ion was computer-selected), and the collision energy was *automatically* adjusted according to its charge state and *m/z* value. The precursor ion was *m/z* 666.266(2+).

From the low mass region, one could predict the presence of Ser, Asn, Asp, His, Phe, and Tyr *immonium* ions at *m/z* 60, 87, 88, 110, 120, and 136, respectively. The *immonium* ion for Asn *m/z* 87 also may indicate an Arg residue, but as we will see, this particular peptide did not contain that amino acid. The next step is to determine the C-terminus. Because this peptide belonged to a tryptic digest, one would expect to see $\mathbf{y}_1$ at *m/z* 147 or 175, indicating C-terminal Lys or Arg, respectively. This CID spectrum does not contain either. However, there is an ion at *m/z* 156.08 instead suggesting a *C-terminal* His. We may be dealing with the *C-terminus* of the protein or with the result of a nonspecific *cleavage*. Fortunately His is also a basic *amino acid*, preferentially retaining the charge; thus, the fragmentation pattern is similar to that displayed by ''*real*'' tryptic peptides. The abundant higher mass ions, *m/z* 1217.55, 1160.54, 1073.44, 910.44, 763.32, 634.25, and 505.24 represent y fragments and reveal mass differences as follows: 57, 87, 163, 147, 129, and 129 Da, corresponding to a GSYFEE sequence. Remember, we are reading the sequence backward from the y fragments! Fragments $\mathbf{y}_7$ – $\mathbf{y}_{10}$ were detected also as doubly charged ions; the instrument afforded more than sufficient resolution for charge state determination from the isotope spacing. (The potential presence of multiply charged fragment ions complicates data interpretation when working with low-resolution CID data of multiply charged ions.) The identity of the *N-terminal* amino acid(s) can be determined from the mass difference between the MH^+ +1 value and the mass of the highest y ion (see Table II). From this calculation, the missing b ion should be at *m/z* 115. If there were two Gly residues at the N-terminus indeed, we should observe a b fragment at this mass. However, if the N-terminus is an Asn residue (identical elemental composition to two Glys), we will not detect $\mathbf{b}_1$ because it is not stable. The first N-terminal ions, *m/z* 172, 144, and 127 (i.e., $\mathbf{b}_2$, $\mathbf{a}_2$, $\mathbf{a}_2$ NH_3), indicate that the latter is the case and our working sequence by now is NGSYFEE. . .H. The lowest **y** fragment identified (*m/z* 505) is very abundant, suggesting that it was formed via a cleavage at the N-terminus of a Pro residue. Indeed, there is a fragment at *m/z* 408.19 that mass is

97 Da (Pro residue weight) lower. Thus, the mass still unaccounted for is 252 Da (y_i-y_1 = 408-156). From the immonium ions, we established that the sequence should contain an aspartic acid that would account for 115 Da and the leftover 137 Da corresponds to a second His. Thus, the options for the C-terminal sequence are PHDH or PDHH. The expected y_2 fragment for the first variation should be at *m/z* 271. The presence of Pro residues usually leads to the formation of abundant internal ions too. PH should be detected at *m/z* 235, and the *carbonyl* loss from this fragment would yield an ion at *m/z* 207. All these ions were detected in the *spectrum*. The other variant should yield a y_2 fragment at *m/z* 293 and internal fragments at m/z 213 and 185. From these three only *m/z* 213.09 was detect ed, which could also repres ent a_3 H_2O. Thus, the full working sequence is NGSYFEEPHDH. To confirm this assignment, most of the remaining ions have to be assigned . The spectrum contains an extensive **b** ion series: at *m/z* 172, 259, 422, 569, 698, and 827 (b_2-b_7), accompanied by satellite ions formed via - 28 (CO) or multiple 18 (H_2O) losses. There are four residues in the sequence that may undergo the latter neutral loss. So, the **y** ions also feature satellite ions due to this process. Quite a few inte rnal ions were also detected in this spectrum; some of them are six to seven residues long, a feature frequently observed in QSTARCID experiments.

PROTEIN IDENTIFICATION USING MS/MS DATA

For protein identification, a handful of sequence ions, toget her with the precursor ion, may be sufficient. Thus, PSD analysis and low- and high - energy CID experiments may be and have been used for this purpose. Software packages provided by the MS instrument manufacturers and by different research groups and software companies are available and can be used for data interpretation. Higher mass accuracy leads to more reliable protein identification even if the MS /MS analysis yields only a few ions.

The importance of peptide sequence analysis in *proteomics* research. The unfractionated in-gel tryptic digest of a yeast protein yielded only a few peptide masses by MALDI TOF MS analysis. Although the yeast genome is known, a database search with these masses specifying a mass accuracy of 150 ppm (*external calibration*) proved inconclusive. (For all the database searches discussed in this chapter, *ProteinPro spector* was used.) The ion at *m/z* 1596.63 was subjected to MALDI CID analysis performed on an ABI 4700 TOF-

TOF *tandem* mass spectrometer. One of the most abundant fragments in its CID spectrum is an ion that corresponds to MH^+-64 Da, indicating that the *peptide* contains a Met *sulfoxide*. Such peptides can be identified by using the MS-Tag feature of *ProteinProspector* in *homology* mode. However, this time this search was not successful. Manual interpretation of the CID data revealed a C-terminal YAAYM(O) FK sequence. An MS-Pattern search with this sequence identified the protein unambiguously as *glyceraldehyde 3-phosphate dehydrogenase*. The missing N-terminal amino acids were PFITND, and the peptide was formed via an Asp - Pro cleavage, which is not an expected hydrolysis site for *trypsin*. Only four of the peptides detected proved to correspond to predicted *tryptic* products for this protein: *m/z* 8 11.41, 1297.63, 1752.79, and 2591.30.

In addition to *nonspecific protease* cleavages and coval ent modifi cations, CID an alysis of the components in unfractionated digests may be complicated by the presence of peptides of identica l or similar molecular mass. The resolution of precursor ion selection has to be considered. While the mass window for precursor ion selection is about ± 5% for PSD e xperiments, the MALDI - TOF - TOF uses an approximately 5–10 Da wide window, triple quadrupoles, hybrid instruments, and ion traps usually operate with a 2–3 Da wide wind ow, and four - sector instruments allow monoiso topic precursor selection . When analysing *unfractionated* digests, it frequently happens that two or more different peptides are fragmented and analysed simultaneously in a single *experiment*.

The electrospray low - energy CID spectrum of precursor ion at *m/z* 724.34 (2+). An unfractionated protein digest was analysed using *nanospray* sample introduction on a QqTOF *tandem mass spectrometer*. The analysis yielded a few high -quality CID spectra suitable for protein identification. The theoretical *tryptic* digest of the protein suggested this particular spectrum might correspond to an oxidised met *hionine*- containing peptide, ITSPLM(O)EPSSI K (MH + 14 47.67), which was confirmed by the CID data. After "*filtering*" out the ions that were expected for this sequence, several abundant fragments remained unaccounted for and readily revealed an *N- termina* l sequence of PFGVALLF. This sequence belongs to the same protein, corresponding to a peptide, PFGVALLFGGVDEK formed by an Arg - *Procleavage*, giving a *protonated monoisotopic molecular* weight of 1448.78. The isotope peaks in the molecular ion cluster were indeed slightly higher than it would be expected for peptides in this mass range. However, isotoperatios may not be reliable as multiply charged peptide ions in unfractionated in-gel digests may not raise significantly above the noise level, and overlapping precursor ions may be masked by other more abundant *species*, possibly of different charge states. Overlapping isotope clusters do occur even in LC/MS experiments, especially when very complex mixtures are analysed.

The MALDI low-energy CID of *m/z* 960.4 in a tryptic digest of a human heat shock protein. These data were also acquired on a QqTOF tandem mass spectrometer. In this case, the isotope pattern of the precursor ion clearly indicates that a mixture was analysed. Such an *isotope* pattern frequently indicates the *hydrolysis* of Asn or Gln residues, either as a common *posttranslational* modification or as potential byproduct of the *isolation/digestion/storage*, giving a partial 1 Da mass increase. However, in this protein, the components were two *unrelated peptides* and the presence of both could be confirmed unambiguously.

ALTERING THE DISSOCIATION PROCESSES BY CHEMICAL DERIVATISATION

De novo sequencing of peptides and protein identification would be accomplished much faster, and at higher sensitivity if one could simplify peptide fragmentation, exclusively generating a single ion series. Of the MS/MS techniques discussed in this chapter, only ECD and ETD fulfill this description. However, it has been reported that generating fixed charges in peptides by chemical derivatisation (i.e., introducing positively or negatively charged groups at a terminal position) will lead to an altered generally more simplified fragmentation pattern. First Johnson showed that peptides that featured a Lys residue converted to a 2,4,6-*trimethylpyridinium* cation exhibited a series of N-terminal or C-terminal ions depending on where the charged residue was located. A series of other derivatisations followed, and it has been established that a fixed positive charge is beneficial for controlling the peptide fragmentation in high-energy CID. However, most of these derivatisations cannot be performed reproducibly at the picomole level or below, may lead to side reactions, and may require purificati on steps . It was repo rted that one of the best deriv atives, "C5Q," an N- termina l trimet *hylammonium* group with a five - carbon linker (i.e., $[CH_3]_3N^+ [CH_2]_5CO$ modi fication), could be introduced in 10 min, at the femtomole level and the derivatised peptide was readily detected by MALDI. However, to generate only N - terminal fragment ions, modification of Lys and Arg side chains was recommended to reduce their basicity. Earlier, Burlet et al. reported "that the form ation of N - or C - terminal precharged derivatives is detrimental to the formation of sequence - specific product ions following low-energy collisional activation.'' In addition, they found that ''protonation of pre-charged derivatives (yieldin g doubl y charged ions) restores favorable fragment ation properties.'' This effect was attributed to the additional proton being able to move freely, occupy any site, and induce charge -directed fragmentation. This effect also explains the ''*y - ladder*'' formation of doubly charged tryptic peptides, especially with C - termina l Arg residues. The same group discovered that the introduction of negatively charged residues by Cys oxidation to cysteic acid in Arg- ended tryptic peptides also promoted uniform fragment ation (i.e., comprehensive *y-ion series* form ation even in singly charged ions) . This finding was consistent with their *hypothesis* ''that increased heterog eneity (with respect to localisation of charge) of the protonated peptide precursor ion population is beneficial to the generation of high yield productions via several charge-directed, low energy fragmentation *pathways*''. Based on these observations, a new higher sensitivity derivatisation was developed in which a sulfonic acid group was added to the *N-termini* of the peptides. This derivatisation with 2- *sulfoben zoic acid cyclic anhydride* is simple and efficient and may be performed directly on target. By donating a *proton*, this acidic group promotes the charge - directed fragmentation of the peptide bonds and, with a basic amino acid such as Arg at the *C- terminus*, yields y-ions almost exclusively.

MS/MS ANALYSIS OF COVALENTLY MODIFIED PEPTIDES

MS can detect the presence of even unanticipated covalent modifications and can provide information on their structure and location. This is one of its greatest advantages over other protein *analytical* techniques. The presence of covalent modifications is usually detected on the *peptide* level in the form of masses that do not match the predicted digestion products. Obviously the covalent modifications can be the results of in vivo processes such as post *translational* or *xenobiotic* modifications, as well as the products of intentional labeling or *fortuitous* chemical reactions. How ever, mass measurement alone does not provide *unambiguous* information on the structure and the exact site of the covalent modification. To prove the structure, suspected peptide fragmentation information has to be obtained. Since for any known sequence the masses of the expected fragment ions can be calculated , the presence of covalent modifications should be detected by the observed mass shifts between the predicted and observed fragment masses. Some structures indeed are very stable upon *collisional* activation, and the site of the modification can be established readily from CID data. For example, acety l groups on the N-terminus or on the E- amino group of a Lys residue will not fall off during any MS/MS experiment. Acetylation is indicated by 42Da shif ts in all the fragment masses that contain the modified residue. In addition, when the N-terminus is modified, the b_1 ion can be detected because it is stable with acyl modification. When the Lys side chain is modified, immonium ions at *m/z* 126.092 and 143.119—much weaker—can be detected.

In addition to acetylation, the most frequently observed stable modifications are the *carboxymethylation*, *carbamidomethylation*, and *acrylamide* modification of Cys side chains, because these modifications are routinely used to block the –SH groups. These derivatisations result in 58-, 57-, and 71-Da mass increase of the Cys-containing fragments, respectively. *Vinyl pyridine* also can be used for *sulfhydryl* protection. This modification is also stable in that one can observe the expected mass shift and the modified *immonium ion*; however, because of the basic structure in it, *S-ethyl-pyridyl* Cys-containing peptides will also display an abundant and characteristic fragment formed via fragmentation of the protecting group: *m/z* 106.066 (CH_2=CH-$C_5H_4NH^+$). The fact that covalent modifications yield abundant fragment ions can be beneficial. If the produced fragment masses are unique enough, they may serve as diagnostic markers and aid the identification of modified molecules. This approach has been used, for example, in the localisation of *glycopeptides*, *xenobiotically* modified species, and ICAT-modified molecules. However, if the modifying group is too successful in retaining the charge at the expense of the other parts of the peptide, it may prevent the assignment of the modification site or even obtaining just sequence confirmation. Charged modifications may interfere with the fragmentation in other ways too. As was presented earlier in this chapter, introducing a positive or negative charge may lead to simplified but definitely altered fragmentation pattern. *Trimethylation* of amino groups is a known posttranslational modification contributing 42 Da to the MH^+ value. The mass difference between acetylation and trimethylation is 36 mmu. However, *trimethylation* differs from acetylation in other aspects, too. It also alters the fragmentation pattern. For example, one cannot observe a shifted *immonium* ion for a trimethylated Lys residue or b1 fragment if the *N-terminus* is modified. Instead *trimethyl*-modified

Lys residues tend to lose 59 Da, that is, $N(CH_3)_3$, as was reported for MALDI analysis and shown for low-energy CID in this chapter. In addition, in high-energy CID experiments, the presence of such a modification may lead to unexpected d and w ion formation. For example, normally a *tryptic peptide* with an Arg residue at its *C-terminus* will yield only w ions. However, the presence of a trimethyl-lysine close to the *N-terminus* will trigger d ion formation.

Characteristic neutral losses like the one mentioned earlier also may serve as diagnostic markers. The 64 Da loss from *methionine-sulfoxide* has been mentioned already. Interestingly, posttranslational modifications of Ser and Thr residues, namely, phosphorylation, *O-glycosylation*, and *sulfation*, also display such preferred fragmentation.

Phosphorylation has been studied extensively by MS. Depending on the site of *phosphorylation*, the modified peptides behave very differently. *Tyr-phosphorylated* peptides retain the modification under most MS and MS-MS conditions. The modified *Tyr* displays an appropriately shifted immonium ion at *m/z* 216, and the site of modification also can be established *unambiguously* from the modified fragment ions present ; 80 Da losses have been reported only in MALDIPSD experiments. At the other extreme, *Hisphosphorylated* peptides undergo extensive fragmentation even under MS conditions. Ser- and Thr - modified peptides are stable during MS acquisition, but the favoured fragmentation step upon collisional activation is the β-elimination of *phosphoric acid* (H_3PO_4) (i.e., 98-Da loss from the molecular ion). Based on this observation, *phosphorylation* sites are regularly assigned based on the presence of ions that are assumed to be the products of such β-elimination. For example, this is the way Mascot assigns phosphorylation sites. Unfortunately, hydroxy amino acids, as well as Glu and Asp residues, readily lose water upon the fragmentation process from all kinds of sequence ions. Such a fragment derived from an unmodified ion will have the same structure (and mass) as the product of phosphoric acid β-elimination from a modified species. Most of the *phosphopeptide* analysis is done online, usually in a data-dependent manner, when the collision energy is set automatically. Although this approach serves quite well the protein identification studies may not be optimal for some post-translational modification assignments. Figure elsewhere in this chapter shows the CID data of a phosphopeptide acquired from its more abundant, 3+ ion (upper panel) automatically and from the less abundant doubly charged ion, after some manual adjustments of the collision energy. The β-elimination products— both due to water or phosphoric acid losses—are very abundant in the spectrum that was taken under preset conditions suitable for the "*meaningful*" fragmentation of most triply charged ions. From these data—when only the β-elimination products are considered—one easily could assume that the Ser residue is modified because every single N-terminal ion detected, starting from $\mathbf{a}_3$, carries the telltale double bond. However, $\mathbf{y}_4$ shifted by 80 Da is the ion that reveals the real *phosphorylation* site. The CID spectrum of the doubly charged ion displays intact $\mathbf{b}_5$ and $\mathbf{b}_6$ ions, indicating that the first six residues are not modified, whereas $\mathbf{y}_3$ shows the presence of modification. Interestingly, the elimination of water from $\mathbf{b}_3$ still produces as abundant an ion as that derived from the *phosphoric acid* loss from modified $\mathbf{b}_7$ and $\mathbf{b}_8$.

In weaker spectra, or under less fortunate circumstances, this preferential fragmentation (i.e., the lack of modified fragments) may prevent the site assignment. However, because this fragmentation also can be *triggered* in the source during MS acquisition even under relatively mild conditions, it may be taken advantage of. The product of β-elimination can be selected as precursor ion and the dehydro-alanine, formed from the modified Ser or the *dehydro-butyric acid*, product of β-elimination from Thr readily can be identified from the 69 and 83 Da residue weights, respectively. Similarly, products of the β-elimination generated in MS/MS experiments in ion traps can be subjected to subsequent MS^3 analyses. The β-elimination also can be accomplished chemically before the MS analysis. Sulfated or glycosylated Ser and Thr residues are also susceptible to chemical β-elimination at about the same extent as their *phosphorylated* counterparts. However, the favored fragmentation step in the CID fragmentation of sulfo- and O-linked *glycopeptides* is a rearrangement reaction that eliminates the modification from the *amino acid* side chain without a trace. Although some of the *glycopeptide* fragments may retain the sugar, the *sulfate* is always eliminated completely. Thus, while the site of *glycosylation* may be determined from CID data, CID-based site assignment is impossible for *sulfopeptides*: their fragmentation pattern is identical to that of the unmodified peptides. The typical low-energy CID fragmentation of peptides modified by O-linked carbohydrates. The oxonium ion of the GlcNAc residue (*m/z* 204) and related *carbohydrate* fragments dominate the low mass region. Peptide fragments that underwent the rearrangement reaction and lost the modifying sugar are usually more abundant than the glycosylated fragments. Still, there is sufficient information in the spectrum to establish Ser-313 (position 11 in the peptide shown) as the site of modification. The side chains usually do not fragment in ECD and ETD experiments.

Thus, the sites of the aforementioned covalent modifications could be assigned easier from such analyses than from CID data. In addition, massaccuracyaffordedbyanFT-MSinstrumentmayfacilitatedifferentiation between isobaric *phosphorylation* and *sulfation*.

8

BIOINFORMATIC METHODS

Just as the study of individual genes has been transformed into the high-throughput and highly context-centered study of whole *genomes*, so has the study of individual proteins been enormously extended through the development of the *multifaceted* field of *proteomics*. Thus, although detailed studies of *enzyme mechanisms* continue to provide important insights into how *biological* systems function at the molecular level, larger scale studies of how proteins behave in pathways contribute new views of *biological chemistry* in the *dynamic* context of the living organism. And although critical information regarding the precise roles of *phosphorylation/dephosphorylation* of specific signaling molecules will continue to be important for understanding individual steps in these networks, larger scale studies of posttranslational processing across entire *proteomes* are being undertaken to understand the role of *phosphorylation* in the overall *dynamic* process.

The new technologies in mass *spectrometric* analysis of peptides and proteins necessary to accommodate *proteomics-scale analyses* require, in turn, concomitant development of informatics technologies suitable for very large-scale data handling and analysis. These technologies represent two primary streams of development: data *handling/storage/retrieval*, as represented by mass spectrometry (MS)–specific laboratory information management systems (LIMS), and data analysis. Until recently, both types of bioinformatic infrastructure have been developed solely by the MS community, tied intimately and specifically to the development of the *physical technologies* themselves. However, *information technology* (IT) experts, programmers, statisticians, and computational biologists are now found in many cutting-edge MS laboratories. This type of intimate collaboration between mass *spectrometrists* and *bioinformatics* professionals will be necessary for the development of the *sophisticated* next-generation *bioinformatic* tools.

MASS SPECTROMETRIC ANALYSIS

The analysis of underivatised peptides and small proteins by MS became feasible with the development of soft ionisation methods, including field desorption, direct *chemical ionisation*, fast atom *bombardment* (FAB), and liquid secondary ion MS. More recently, the introduction of electrospray ionisation (ESI) and *matrix-assisted* laser desorption ionisation (MALDI) enabled the *routine* and highly sensitive detection of isolated gas-phase ions of intact proteins of virtually any size. Despite this capability, relatively little use has been made of the analysis of intact proteins for protein identification. The mass spectrum of a protein typically gives only a single piece of information, its molecular weight. Furthermore, unless this measurement is made in an *ultrahigh* resolution instrument such as a Fourier transform (FT) ion cyclotron resonance (ICR) mass spectrometer, only for the smallest proteins will the *isotopic* components of the *molecular* ion be resolvable, and the measurement of molecular weight will not be sufficiently accurate to distinguish between hundreds or even thousands of possible *isobaric* species. In addition, the molecular mass of a protein will frequently differ from that predicted by the gene sequence, which can be attributable to events such as alternative splicing, *posttranslational* processing of the expressed protein or subsequent chemical modifications that can either add or remove mass, or even errors in the databases. Thus, for an identified protein, the molecular mass (or masses for a *heterogeneously* modified protein) may not be very revealing.

Analyses that start from the intact protein molecular ion have come to be known as "*top-down*." Although the collision-induced dissociation (CID) techniques that readily give sequence information for peptide ions are generally not applicable to species as large as intact protein ions, the fragmentation mechanisms of *electron-capture* dissociation (ECD) and electron-transfer dissociation (ETD) make direct protein sequencing a realistic possibility (Horn et al., 2000). Here, the attachment of a low thermal energy *electron* to one of the positive charges in a multiply charged ion with a recombination energy of about 6 eV gives a radical ion that will rapidly fragment before energy *equilibration* occurs. Such species are classified as distonic ions because the charge and radical sites are distinct from each other. As in the well-known radical cation reactions of species formed by *electron* impact, alpha-cleavage occurs not at the site of the charge but one bond removed, giving *predominant* c- and z-ions for cleavage between the amide nitrogen and the α-carbon. This differs from CID techniques that predominantly produce b- and y-type ions.

The alternative ''*bottom-up*'' approach has been the major focus for protein identification during the last decade, based on the demonstration that digestion of a protein with a protease such as trypsin and the analysis of the resulting peptide mixture can provide sufficient information to identify it in a protein database. The specificity of the protease is known, so the molecular weights of the resulting peptides can be predicted for any protein sequence. Even when the peptide masses are measured in an instrument of relatively modest performance, proteins may be identified from the detection of a relatively small subset of the possible peptides, using the technique referred to as peptide mass mapping or mass *fingerprinting*. As genomic databases started to grow more rapidly, it was realised that the criteria for protein identification would need to be more stringent and it became necessary to improve the accuracy of mass measurement. For routine operation, this was achieved by moving from linear to *reflectron* MALDI TOF instruments with *delayed extraction*.

Whereas peptide mass fingerprinting is able to identify a purified protein from its digest, the technique's performance becomes unreliable if a mixture of components is present in the sample. For more confident and reliable identification, fragmentation *spectra* produced by tandem MS can be matched against sequence tags predicted for all proteins in a database. As originally formulated, mass tags represent a series of fragment ions that can be attributed to a *coherent* sequence of *amino acids* that is a subset of the sequence of the peptide under study. Further developments that are usually grouped with mass tag *searches* are actually based on comparisons between the experimentally observed fragment ions and all predicted fragments for all *hypothetical peptides* of the appropriate parent mass, based on known fragmentation rules. If the peptide of interest is not present in the database, de novo peptide sequencing can be undertaken, again based on known rules for peptide fragmentation. This is most important when a significant proportion of peptides diverge from those predicted due to errors or incompleteness of databases, *discrepancies* between genomic sequence and processed or *posttranslationally* modified proteins, species differences, and *nonspecific* enzyme cleavages. Techniques have been developed to identify proteins based on sequence homologies to related proteins or equivalent proteins in different species, although these are time consuming and computationally intensive compared with straightforward searching based on sequence tags.

Good mass accuracy not only increases the reliability of database searching by limiting the possible compositions of peptides for any given mass but also reduces the search times. The quality of mass spectral information that has been employed for proteomic applications has ranged from low-resolution MALDI-TOF spectra in which mass accuracy of ±1 Da was routine to FT-ICR measurements achieving better than 1 ppm *accuracy*. Routine operation in a DE MALDI-TOF with reflectron gives better than 50-ppm mass accuracy, and significantly better (~10 ppm) may be achieved with careful internal calibration. ESI sources are commonly used on linear quadrupole instruments or *quadrupole* ions traps that typically give fairly modest performance in terms of resolution and mass accuracy, although the combination of a quadrupole mass selector and quadrupole collision cell with orthogonal acceleration (QqoaTOF) gives high resolution (~10,000) and perhaps 10-ppm mass accuracy if well calibrated. FT-ICR instrumentation can routinely achieve 2-ppm mass accuracy or better, which in many cases is sufficient to specify the elemental composition of the peptide.

Sequence ion information from a MALDI-TOF instrument using post-source decay (PSD) is useful, although it is painstaking to collect and the accuracy of mass measurements is relatively low, so this is generally employed only when no alternative tandem MS instrument is available in a lab. Higher quality CID spectra can be obtained by MALDI-TOF=TOF, ion trap, or QqTOF *geometry* instruments employing either MALDI or ESI ionisation.

Existing Tools

Several web sites offer free access to web-based database-searching programs for peptide mass fingerprinting and the identification of sequence tags, all of which provide other tools as well, such as prediction of *enzyme* digests and prediction of CID fragments. An alternative to Internet access is to purchase a license to have the programs resident in-house, and most *mass spectrometer* vendors supply software and software licenses with the purchase of an instrument. Having an in-house system provides advantages of data security and the ability to use private databases or databases configured for specific tasks. It gives the user the option of employing a more powerful computer or installing software on a computer cluster to speed up search times.

Some packages available for online searching include tools on the ExPASy *proteomics* server provided by the Swiss Institute of *Bioinformatics*, Mascot from Matrix Science, Peptide Search from the European *Molecular Biology* Laboratory, *ProteinProspector* from the University of California, San Francisco, or distributed by Applied Biosystems, Xtandem! (as part of the Global Proteome Organisation machine), and Prowl from Rocke feller University, or related Knexus software. Other software distributed by instrument manufacturers include Sequest from Thermo Finnigan and Spectrum Mill from Agilent Technologies, and other manufacturers produce their own software analysis packages. Most of these packages contain a number of tools, including programs for identifying proteins from a list of peptide masses (mass mapping or *fingerprinting*), fragment masses (mass tags), or single or multiple sections of sequence tags, which could come from de novo *interpretation* of CID spectra or from other sources.

With the availability of non *ergodic* fragmentation mechanisms that can efficiently fragment small intact proteins, searching tools need to be developed to analyse these data. This type of data can be conceptually analysed using software for analysing peptide fragmentation such as MS Tag (see later discussion in this chapter). However, specific software for this task is being developed, the first of which, "*Prosight* PTM," is designed for identifying modification sites from intact protein fragmentation.

The major instrument types used for acquisition of tandem MS data are ion trap instruments and QqTOF *geometry* instruments. Data acquired on an ion trap instrument generally have a lower information content than QqTOF data, as the lower resolution leads to lack of charge state determination of parent and fragment ions, the lower mass accuracy requires wider mass tolerances, and the absence of the low mass region loses further information. Hence, *discrimination* between correct and incorrect matches is more difficult with this type of data. A popular search engine for analysis of ion trap data is Sequest, which was specifically designed for this type of data. Several research groups have now developed tools that *reinterrogate Sequest* results using multiple factors about the search result to produce more reliable *peptide* and *protein* summaries.

The higher information content of data acquired on QqTOF

geometry instruments allows better discrimination between correct and incorrect answers, but false assign ments are still made. One has to remember that the output from the searc h resul t is a "*best fit*" to seque nces in a database, with an attempt to apply a significance threshold. Al most all database searches result in incorrect second best matches for particular spectra, which would pass all scoring thresholds and would be reported correct if the best matching sequence was not in the database, or if the correct peptide bears an unaccounted for modification. Thus, there are nearly always going to be incorrect *peptide* assignments in noncurated analyses.

With out a community standard for protein identification, a significant issue within the proteomics field is the difficulty of assessing and comparing results acquired on different instrument platforms and/or analysed using different search *algorithms*. This creates a significant challenge for journal reviewers and editors with respect to assessing whether submitted results are of sufficient quality/ confidence for publication. In response to this, the journal Molecular and Cellular *Proteomics* published some initial guidelines for data submission. Although these are not expected to be rigid or in an y way comprehensive, the journal 's intention is to provoke discussion about the subject so some rules can be applied within the community as a whole.

Many of these problems are caused because raw MS data can currently be stored only in manufacturer specific data formats, which are not compatible with each other. As a result, it is difficult to develop generic tools for proteomic data analysis and to accurately assess how different search engines are performing. Several groups are aiming to solve this problem. In a major initiative at the Institute for Systems Biology in *Seattle*, open source tools for analysis of MS data acquired on any instrument platform are being developed, allowing direct comparison of data acquired on different inst ruments. This project includes software to convert data from any *mass spectrometer* into a common data format, tools for *reassessing* and improving reliability of search results, and tools for analysis of samples labeled with isotopic *quantitation* reagents. The Manitoba Proteomics Center also is making available open source software for capture and analysis of raw data acquired on many instru ment platforms. The Human Proteome (HUPO) initiative is also attempting to promote a common data format. These are all noble efforts; it remains to be seen whether any of these will be accepted on a large scale by the *proteomics community*.

ANATOMY OF A WEB-BASED PACKAGE: PROTEINPROSPECTOR

As previously stated, a large number of tools are a vailable on the web for an alysis of *proteomic data*. *Protein Prospector* is one of the more complete packages of programs for database analysis, and we describe these programs further to exemplify the types of tools available.

ProteinProspector, a suite of programs described as providing "*tools* for mining sequence databases in conjunction with *mass proteometry* experiments," was created by Karl Clauser and Peter Baker at the University of California, San Francisco. The software has a web-based interface, as well as a mirror site at the Ludwig Institute in London. All Protein Prospector programs are written in the C++ programming language. They receive their input according to the standard common *gateway interface* (CGI) *protocol*, and initially produced output in HTML format, although XML is a recent introduction to the current version available over the web at the time of this writing (Version 4.0.5). Output of the programs can be readily parsed for downstream organisation and storage in a relational database. The programs also run from the command line, which lends itself to *automation*.

The current web version of Protein Prospector contains the following programs:

Sequence Database Search Programs

- *MS-Fit*. A peptide mass fingerprinting tool that matches molecular weights of peptides observed from the enzymatic digest of a protein to the peptides predicted for each protein sequence in a database.
- *MS-Tag*. Identifies proteins based on comparing a list of fragment ion masses from a CID spectrum of a peptide to theoretical fragment ions for each predicted peptide of the same molecular mass in the database produced by a given enzyme cleavage specificity.
- *MS-Pattern*. A database-searching tool that performs text-based searches of amino acid sequence (with or without peptide mass filtering) against sequence databases.
- *MS-Homology*. A similar tool to MS-Pattern, it can take CID MS data as any combination of amino acid tags and masses and report peptides identical or homologous in the database.

It can search data from multiple CID spectra and require a minimum number of peptides to match to the same protein in order for it to be reported. It is designed for identifying *homologous* proteins in the database, and because of its ability to search with multiple sequences, it is possible to identify remote homology.

- *MS - Seq*. An alternative to MS - Tag based on "Peptid eSearch".
- *MS - Bridge*. This identifies inte rmolecul ar or intramolecular disulfide- linked peptides and can be used to identify chemical cross -linked species with user - definable structures.

Peptide /Protein Utility Programs

- *MS - Digest*. Performs an in silico enzymatic digestion of a protein sequence and reports the mass of each predicted peptide. Possible inputs include the MS - Digest index number produced by MS - Fit, MS - Tag, MS - Seq, or MS - Pattern, the accession number from a database, or a user - supplied sequence.
- *MS - Product*. Calculates all fragment ions resulting from fragmentation of a peptide. The fragment ion types reported are defined by specifying the instrument type used for acquisition of the data or can be selected by the user.
- *MS - Isotope*. Calcul ates stable isotope profiles of peptides (or other organic molecules) based on peak shape, resolution, and charge. It can plot two or more over lapping profiles, including peptides of different charges but similar *m/z* values. In addition to specifying an elemental composition or a peptide sequence, an " *average mass* " can be specified to give a "*typical*" profile for a peptide at an y given mass and charge combination.
- *MS - Comp*. Calculates possible *amino acid* compositions of a peptide based on a parent or fragment *m/z* and partial composition from the *immonium ions* or other data contained in a CID spectrum. It is possible to specify various alternative ion types.
- *DB - Stat*. Calculates useful statistics for any specified database.

A variety of databases are available through the Internet for searching proteomic data. These range from large essentially *non-curated* protein sequence *repositories* to smaller highly annotated databases to *theoretical translations* of EST sequences.

The most highly curated pro tein databas e a vailable is SwissProt. This is a non - redundant database, in which multiple sequences for a given protein are collapsed into one entry. It contains extensive integration and links to other reference databases including information about protein domains and structure and protein families. As this database is very highly curated, new entries into the database are relatively slow to arrive. Hence, a less curated, but more rapidly expanding offshoot TrEMBL allows quicker entry of a sequence into the database. This is still annotated, but the curation is largely autom ated based on similarity to pro teins in the SwissProt database. These two databases, along with another annotated database, PIR-PSD, have been combined into a new set of databases under the UniProt consortium. UniProt contains additional annotation including geneontology annotation, which attempts to assign function, subcellular location, and processes in which a protein is involved.

The largest and most comprehensive database is the NCBI's Entrez protein database, which is formed by the combination of a number of databases including Genpept, which is a translation of the Genbank nucleic acid sequence database, an annotated collection of all publicly available nucleotide. It is part of the International Nucleotide Sequence Database Collaboration, which also includes the DNA Data-Bank of Japan (DDBJ) and that of the European Molecular Biology Laboratory (EMBL). These three organisations exchange data regularly. Many of the proteins in Genpept are also included in other well curated databases such as SwissProt, although SwissProt accession numbers are not always accessible through Genpept annotations pages (and vice versa). Many genes and their protein products are now represented in Entrez by "Reference Sequences." This RefSeq collection was developed to provide a comprehensive, integrated, non redundant set of sequences, including genomic DNA, transcript (RNA), and protein products, for major research organisms. Genpept contains a large amount of redundancy, and entries have variable levels of annotation. Hence, some entries (e.g., those from SwissProt) are highly annotated, whereas other entries may be no more than hypothetical protein sequences based on gene sequence. For more detail, a review of protein database options has been published.

The search programs allow effective interrogation of any FASTA formatted database containing either protein or DNA sequences. The

FASTA format was chosen primarily because of its universality, brevity, and expected ease with which database files could be shared on the same computer with other programs for sequence analysis.

When a database is first downloaded to a given search engine, most search engines run a program that analyzes the entries in the database, assigns them an index number, indexes them according to species, and may calculate protein molecular masses and theoretical pI. This permits restricting the proteins searched on the basis of species, protein molecular weight, or pI. It also allows users to add proteins to databases or create a new database containing user defined proteins. Searches performed on these smaller databases restricted by species or mass are generally very much faster than searches performed on complete databases.

To allow users to take full advantage of the range of data available from different types of experiments, the search programs in the Protein Prospector package are structured so that the results from a search with one program can be input directly into another program. This allows a user to effectively combine information from several analyses in silico for maximum *discriminating* power in defining protein identity. The search programs are also *homology* tolerant, and homology matching is implemented in a general manner so it makes little differenc whether a mismatched amino acid is the result of a substitution by one of the standard amino acids or a chemical, *posttranslational*, or *xenobiotic* modification. All programs based on ion fragmentation incorporate the most current knowledge of gas phase ion fragmentation. However, MS=MS spectra from newly discovered modifications of peptides may contain previously unknown fragment ion types. Consequently, the programs have been developed so that they need only min or modifications to incorporate new fragment ion types.

Data supplied for searching in the aforementioned programs are input manually in the form of peak lists. However, the increase in robustness and reliability of tandem MS data has led to an explosion in the amount of data being acquired. With the routine acquisition of hundreds (or thousands) of fragmentation spectra in a given LC MS analysis, the manual sub mission of peak lists from each spect rum is impractical. Thus, to analyze this type of data, a search engine must be able to accept as input text files that contain peak lists of all precursor ion masses and fragment ions produced and then to produce a succinct summary of the resulting identifications. Several search engines can use text files of this format as input, although the exact format of the file differs slightly between search engines. A developmental version of ProteinProspector is capable of accepting text files from formats used by other search *engines* such as Mascot. Searchi ng data in this form at is performed essentially by sequential searches of each spectrum with MS Tag or an equivalent program. However, searching in this manner is dramatically faster than the sum of individual searches, as many of the calculations required for individual searches are in common for all searches (such as the in silico digestion of all proteins in the database). Thus, a search of more than 3000 MS=MS spectra against all proteins in the SwissProt database, even allowing for a few peptide modifications, is completed in about 20 min on a single processor desktop PC.

As well as trying to identify unmodified peptides, search engines can also be programmed to identify *biologically* or chemically modified peptides. These modifications can be specified as fixed; for example, it is typical to reduce and *alkylate cysteine* residues with, for example, iodoacetamide before protein digestion, so that all cysteines will be *carbamidomethylated*.

Other modifications can be variable; for example, *methionines* can be *biologically* oxidised, but more often they become oxidised during sample preparation. Therefore, peptides should be searched both as if methionines were unmodified and oxidised. Also, most proteins are acetylated on their *N-terminus*, and if after enzyme cleavage, a *glutamine* becomes an *N-terminal* residue, this can pyrolise to form a *pyro-glutamate* residue. These modifications are generally searched for as a default. Other modifications such as phosphorylation can also be searched for, although these searches do not take previous biological knowledge into consideration and thus will search for any serine, threonine, or tyrosine being modified. This greatly increases search times and leads to many more incorrect answers, so it is generally advisable not to search for such modifications as a default.

Enzyme digests typically employ *trypsin*, but any other enzyme can be selected. Combinations of *enzyme* cleavages can also be specified or data can be searched with no *enzyme* specificity specified for cleavage. The last of these options leads to a high number of false positives and should be used only when there is supporting evidence for such *peptide* products present.

For mass fingerprinting data, it is typical to specify a minimum number of peptide matches for a hit. The order of reporting results is usually by a scoring system, the most common being based on the MOWSE system, which is used by Mascot and MS-Fit in ProteinProspector.

When searching tandem MS data, the user must tell the search engine what type of instrumentation was used to fragment the peptides soit knows which fragment ion types to consider. For example, if fragmentation data were acquired on a MALDI-TOF=TOF instrument, additional high-energy CID fragmentation processes must be considered. Some search engines are unable to use sequence specific rules for fragmentation, but MS-Tag in Protein Prospector uses the following rules for high-energy CID analysis:

- No side-chain fragmentation should be anticipated unless an arginine, lysine, or histidine residue is present within the ion. In general, the basic residue will not itself yield side-chain fragments: for example, C-terminal arginine does not form v- and w-ions, although it may yield a d-fragment if there is another basic amino acid in the peptide.
- No d-ions should be formed by *alanine*, *glycine*, *histidine*, *phenylalanine*, *proline*, *tryptophan*, or *tyrosine*.
- No w-ions should be formed by *alanine*, *glycine*, *histidine*, *phenylalanine*, *tryptophan*, or *tyrosine*.
- Because of the double substitution of the β-carbons in threonine and *isoleucine*, each can yield two w- and two d-fragments.
- No v-ions should be formed by glycine or proline.
- No c-fragments should occur for amino acids that are N-terminally adjacent to proline residues.
- Fragment z (i.e., y-16) cannot be formed by proline residues.
- A neutral loss of NH_3 (-17 Da) from the normal sequence ions and internal fragments is considered for species that contain *arginine*, *asparagine*, *glutamine*, or *lysine*.
- A neutral loss of H_2O (-18 Da) from the normal sequence ions and internal fragments is considered for species that contain *aspartic acid*, *glutamic acid*, *serine*, or *threonine*.
- The first members of the N-terminal ion series are not listed as such. In general b_1 is not stable unless the N-terminus is modified, the a_1-ion is equivalent to an immonium ion, and c_1 is rarely observed.

What do You do When Your Protein is Not in the Database?

Unless proteins share very high sequence identity, their tryptic peptides will not always be conserved across closely related *species* or even among isoforms within a single species. This, combined with the presence of sequence errors in databases, means that identification of proteins, especially from species that are not well represented in databases, may not be achieved using the conventional searching approach described earlier in this chapter. In these cases, some level of de novo sequencing is required. There are two levels to which de novo sequencing can be applied.

In one approach, a short stretch of amino acid "*sequence tag*" is discerned, and this, often combined with the masses of fragments before and after this amino acid tag, can be used to search a database in an error-tolerant fashion using software such as MS-Homology or PeptideSearch.

The second approach for identifying unassigned spectra is for software to attempt to de novo sequence the entire peptide and return a "*best-fit*" sequence without referencing a database. This approach relies on scoring systems to rank the quality of de novo sequence matches to the data. These scoring systems have to be different from those used for database searching, as there is no "*look-up table*" to which to compare the results. A few software packages are available for this type of analysis: Sherenga, Lutefisk, and commercially available PEAKS.

Once a sequence is determined, search engines such as BLAST and FASTA can be used with sequence information obtained by de novo peptide sequencing of several peptides taken from a tryptic digest. However, these searches generally are performed one peptide at a time. In addition, several *idiosyncratic* features of fragmentation data from mass spectrometers are not adequately accommodated for in these search engines. First, sequences from MS results are almost invariably relatively short stretches of amino acid sequence (often <10 amino acids). The statistical significance of sequence matches using these search engines is length dependent, and searches using such short sequences do not produce statistically significant matches, especially when no very close homologue exists in the databases. The fragmentation mechanism of CID generally prevents the formation of a b_1 ion; thus, identification of the N terminal two amino acids is often not possible. Distinguishing between the pairs Ile/Leu and Gln/Lys is also generally not possible, although Ile/Leu can be differentiated using high energy CID, and glycine and lysine can be distinguished between by accurate mass measurement of relevant sequential ion series. Often, a few ions are missing from an ion series in the fragmentation spectrum, leading to ''*gaps*'' in the sequence that cannot be determined.

Hence, MS specific de novo sequencing and *homology* matching search engines have been developed that employ new scoring *matrices* that allow for the possible ambiguities between Ile/Leu and Gln/Lys, can cope with the mass gaps in sequence information, and can use multiple short stretches of sequence.

QUANTITATION OF PROTEINS

Although extensive resources have been applied to the development of MS as a tool for protein identification, only recently has protein quantitation been addressed seriously with respect to more global protein levels. One approach to analysing proteins in complex mixtures is to separate proteins by two dimensional (2D) *polyacrylamide* gel *electrophoresis* (PAGE) before mass *spectrometric* analysis. This technique yields a level of quantitative information based on the *intensity* of staining and is well suited to monitor variations in protein expression in cells or tissues under different sets of conditions. Thus, comparisons of gels enable specific proteins of interest to be identified for more detailed study by MS. However, many researchers have made a move away from the use of (2D) gels with more reliance on other separation methods such as *multidimensional* high performance liquid *chromatography* (HPLC), typically strong cation exchange *chromatography* followed by reversed phase. There is also more emphasis on one dimensional (1D) gels, as 2D gels are not as effective at analysing certain classes of proteins, most notably membrane proteins. However, *quantitation* by staining intensity of 1D gel bands is not reliable for individual protein *quantitation* because bands frequently represent several unseparated proteins.

MS has been used for *quantitation* almost from the time of its invention, even though as a quantitative tool, it has well known deficiencies. For example, the essential conversion of molecules to ions is generally incomplete and the relative *ionisation* efficiencies for different compounds in different circumstances are rarely predictable. Furthermore, the actual numbers of ions released into the gas phase in ESI or MALDI may be highly dependent on the *ionisation* method and other materials present, including other analytes, matrices, and *impurities* arising from purification *protocols*. In addition, the response of the instrument may not be the same for all ions, depending on mass, charge, and polarity.

However, it has long been appreciated that all of these deficiencies can be overcome if an internal standard can be employed with virtually identical *physical* and chemical properties. This is best achieved with an *isotopically* labeled analogue of the compound under study, which will behave identically during *ionisation* but can be differentiated by mass. The replacement of several *hydrogen* atoms by *deuteriums* is generally the least expensive option and has been employed widely and successfully in innumerable studies. The *hydrogen* atoms to be replaced must not be labile under the conditions of the experiment, and there should be a sufficient number of them to separate the natural isotopic clusters for the analyte and the standard. *Quantitation* is then achieved by comparing the relative ion currents and relating them to the amount of the standard added.

Using a similar underlying *philosophy*, methods have been developed to compare protein expression under different conditions, in which one population of cells is grown in a normal expression medium and a second population is in an *isotopically* different environment (e.g., using ^{15}N in place of ^{14}N in the growth media). The cells are then mixed and the proteins extracted for MS. All proteins that are expressed equally should give the same peak *intensities* for the normal and enriched isotope, whereas upregulation or downregulation would be revealed by significant differences in the abundance ratios. The use of isotope depleted material has also been explored in a similar manner. An adaptation of this approach has been successfully applied in which one sample to be compared is grown in media where one amino acid is isotopically labeled, and after mixing samples, peak intensities of peptides containing this amino acid are compared.

If the samples to be compared are not produced by cell culture, which is generally the case for clinical samples or samples derived from tissue, then chemical derivatisation is employed rather than metabolic labeling. Such reagents have specific reactivity for either the N terminus of a peptide or a certain amino acid. The first widely used reagent of this type was described as an *isotope* coded affinity tag (ICAT) reagent. This compound contains a *biotin* tag for affinity purification, a linker that in the light form contains no heavy *isotopes* and that in the heavy form has eight *deuterium* atoms, as well as a group that is reactive toward *cysteine* residues. Thus, two populations of proteins can be differentially labeled, combined, purified over a *streptavidin* column, digested, and the *peptides* analysed by MALDI or ESI MS, giving pairs of peaks separated by 8 Da. A few studies were successful with this reagent, but unfortunately, the large reagent tag was readily fragmented in tandem MS to produce a number of fragment ions not derived from the peptide, and this had a negative impact on the quality of fragmentation spectra. In addition, the slightly less polar heavy deuterated form eluted later in RP HPLC, making quantitation after chromatographic separation problematic. Newer reagents have switched to the use of ^{13}C, ^{15}N, or ^{18}O , which do not produce *chromatographic* differences. Also, cleavable linker regions have been incorporated that allow removal of the majority of the tag after affinity purification and, thus, have minimal side effects on the quality of the fragmentation spectra.

It was envisaged that the ability to selectively enrich only a few peptides from each protein (e.g., cysteine containing peptides) would allow more comprehensive coverage of the proteome in a complex sample. However, this has not been the case, and the reliability of *quantitation* of a protein on the basis of identifying only one peptide from it is *questionable*. Also, there is increased interest in analysing posttranslational modifications on these proteins. The latest commercial set of reagents for protein *quantitation* label the N terminus of peptides and lysine residues. The labeling of all peptides allows better statistics for protein quantitation and is amenable to quantitation of post-translational modifications. In addition, this set of reagents allows the comparison of four samples in one experiment. One drawback is that labeling is not performed until after peptide digestion, and as a result, there is increased potential for introducing variability of the results due to sample handling.

Software for the analysis of these data is limited, and there is no community standard method for analysing this type of data. Applied Biosystems, which markets the ICAT and iTRAQ reagents, has written software called ProQuant, open-source software that is available from the Institute for Systems Biology, and the in-house developmental version of Protein Prospector has software that can perform this type of analysis. Quantitation can be measured based on a number of factors: monoisotopic peak intensity or area or the summed intensity or peak area of the whole *isotope* cluster. *Theoretically*, the comparative peak areas of the whole *isotope* cluster should be most accurate, although there are potential complications with all methods: Mono-isotopic peaks are often weak for high *mass peptides*, but using the whole isotope cluster increases the potential for problems caused by co-eluting peaks of similar mass. In our experience, any of these parameters give similar reliability and standard deviation for results. Complications arise when one or both of the *isotopic pairs* give peaks that barely exceed the signal to noise of the *spectrum*. Without accurate calculation of the background noise level, very inaccurate ratios can be calculated in these circumstances. Hence, when calculating protein quantitation ratios, it is prudent not to use peaks where one or both of a pair of isotopically differentiated peaks are close to the noise level. However, ignoring certain peptide signals will limit the *dynamic range* of the quantitation strategy.

THE CHALLENGE OF AUTOMATION AND HIGH THROUGHPUT

Generally, the promise of high-throughput experimental genomic and proteomic sciences has developed in parallel with approaches to *bioinformatic* analysis of the data using *semiautomated* or fully *automated* routines. As MS-based proteomics efforts gear up to accommodate thousands of samples per day, automation of the entire process including sample preparation, MS, and *informatics* analysis becomes an absolute necessity.

Techniques for automated gel spot identification, cutting, and digestion combined with automated MS and CID analysis by online HPLC with ESI or multiple sample deposition for off-line MALDI are rapidly establishing themselves as routine methods for large-scale protein identification. In so-called data-dependent experiments (DDEs), with HPLC coupled to a tandem *mass spectrometer*, repetitive recording of MS spectra is interleaved with the selection and analysis of peaks for CID *analysis*. The amount of data generated by such experiments running on a continuous basis is overwhelming for manual *interpretation* and automation is essential.

One issue concerning the automated transfer of mass *spectrometric* data to the informatics algorithms is the accuracy of so-called ''*peak-picking*'' algorithms. Such programs convert the instrument-specific and technique-specific raw mass *spectra* into generic lists of *monoisotopic* masses and intensities, usually after some spectral processing to enhance the peaks while reducing electronic and random noise. The mass value for each peak is usually defined by its centroid, determined for a certain fraction of the peak height (i.e., for the upper 50%), which is selected to avoid the inclusion of noise at the leading or trailing edge and to allow some failure to fully resolve adjacent peaks. MALDI-generated ions are assumed to be singly charged, and the challenge is to correctly identify the first peak in each isotopic cluster. For well-resolved isolated peptide ions of m=z values less than approximately 1500 with good signal-to-noise ratios, this is a relatively easy task because the first peak in each cluster (i.e., the monoisotopic peak) is the most abundant. Above m=z 2000, this is no longer true and ions in this region are frequently weaker in intensity. Therefore, the correct identification of the monoisotopic peak becomes progressively more difficult with increasing mass, particularly if it falls below a threshold selected to discriminate between actual peaks and background noise. This problem is compounded by the overlapping isotopic clusters likely to occur in the spectra of peptide mixtures. An *algorithm* has been described that picks *monoisotopic* peaks based on *Poisson* modeling of theoretical isotope patterns compared with the experimental data. For ESI-generated ions, there is the further issue of *identification* of the charge-state (z). the spacing of the *isotopic* peaks on the abscissa is equal to 1=z, the identification of which places certain demands on the resolving power of the mass *analyser*. If the peak spacing can be identified correctly, conversion of the measured m=z to the molecular mass of the peptide is trivial, but for weaker peaks, or if the spectrum is noisy, then software that creates peak lists for database searching will make mistakes. Some instrument types such as ion traps do not produce spectra with distinguishable charge states in full scan spectra. A slower narrower mass range "*zoom scan*" would allow charge-state determination, but for high-throughput analysis of complex mixtures, many researchers decide to skip this extra scan and guess the charge states for a given spectra. For *tryptic peptides*, it is normal to assume a charge state of either 2+ or 3+. The free N-terminus and basic C-terminal residue are both expected to be protonated, and if there is another basic residue in the sequence, such as a *histidine*, this is likely to become *protonated* as well. Peak lists are generated as if there is a doubly and *triply* charged species, then whichever gives the higher, more significant match after database searching is assumed to be the correct charge. This, of course, doubles the number of spectra to be searched and does not always correctly assign the charge state, increasing the number of incorrect identifications.

A major requirement that has arisen as a result of *acquisition* of such large numbers of spectra is the developm ent of scoring systems that attempt to effectively *discriminate* between correct and incorrect answers. It is impractical to manually verify the results from such large data sets. As a result, scoring systems for database results from search engines must be able to report the results at a measurable degree of reliability or with a statistical significance. Search engines such as "*Mascot*" and "*Sonar*" use probability - based scoring systems based on the likeliho od of a given numb er of pe aks in a fragmentation spectrum matching at random. Another approach has been to apply statistical analysis of search results to create a new scoring system. The assumption is made that because the incorrect matches are random, they will form a gaussian distribution of scores that will be lower than the correct matches. Hence, from the database search results, it is possible to identify the distribution of scores for incorrect answers and assign a probability that a match is incorrect. This can be applied at both the peptide and the protein level. The latest version of Protein Prospector is using a similar approach in an attempt to reliably distinguish between correct and incorrect answers. A limitation that should also be acknowledged with statistical-based search results is that these become more reliable with the more data points (spectra) there are to model the scoring. Hence, for the analysis of small data sets, their *effectiveness* and reliability is diminished.

Another method that has been proposed to increase the reliability of results is to search data with multiple search engines and compare the results. The argument goes that as most search engines use slightly different approaches and scoring if a *peptide/protein* is identified by multiple search engines, then it is much more likely to be correct. The developmental version of Protein Prospector allows comparison of multiple searches of a given data set, including comparison of Protein-Prospector and Mascot searches of the same data. These comparisons are also useful for the *development*, *refinement*, and *benchmarking* of search engines.

Although most search engines perform well, there is still room for improvement, because there is additional information within the data that is not taken into account by many search engines. For example, the probability of observing a given fragment ion is peptide sequence specific. An in-depth analysis of peptide fragmentation occurring in the N- and C-terminal to each amino acid demonstrated that certain *cleavages* are highly favoured (e.g., *cleavages* N-terminal to a proline or C-terminal to an aspartic acid). The fragmentation is also affected by the charge state and number of basic residues in the sequence, with singly charged peptides being most influenced by sequence specific fragmentation. Thus, incorporating this information into search *engine scoring* could improve the reliability of results. Other ancillary information can be exploited (e.g., by correlating actual peptide elution times from reverse phase *chromatography* with predicted elution times based on hydrophobicity). Similarly, for peptides purified by cation exchange or isoelectrofocusing in the first dimension, the distribution of peptides between the various fractions can be correl ated with predicted basicity or pI.

Through storage of previous proteomic results in a relational database, several other factors can be used to increase confidence in the results. When a protein is identified in a proteomics experiment, it is usually on the basis of only a few *peptides*. These peptides that are observed from a given protein are generally the same in each *proteomics* experiment. Therefore, a list of "*favoured*" peptides that are observed when identifying a certain protein can be created, and this could be used as a weighting factor if the search results report identification of a *protein*, especially when the identification is on the basis of only one or two peptides. Similarly, rather than comparing a spectrum to a theoretical spectrum created by a search *engine*, one could compare the *spectrum* to a spectrum in which the same peptide was identified before. The Global Proteome Machine Organisation is a public web site where one can search proteomic data with the search engine Xtandem!, and it store s the results from all search es performed on the site. It allows users to compare their results to previous results stored on the site when identifying the same protein in terms of *peptides* identified, the *spectrum* obtained for each peptide, and other proteins other investigators have identified in association with the protein of interest.

Protein identification in simple protein mixtures using MS for *genomes* that are fully known has become a routine exercise. The focuses now are on providing effective methods to describe complex samples in a reliable manner, approaches to fully characterise proteins that are post translationally modified, and the development of techniques that can provide relative and absolute quantitation of components. These approaches will all be driven by a combination of new sample preparation and separation *strategies*, combined with the development of *bioinformatics technologies* to maximise the utilisation of the data acquired.

9

Isolation of Enzyme

Before attempting to extract an *enzyme* from an organism, some preliminary considerations should be made.

For a commercial venture the most important requirement would be to use a source which enabled large amounts of a suitable enzyme to be extracted by some *convenient* procedure: any source which fulfilled this requirement could be chosen. On the other hand, a programme of scientific investigation might require the *extraction* of a given isoenzyme from a specified source, no alternatives being permissible. In the present chapter, therefore, we will discuss *principles* of extraction which are of general application.

First of all, it is essential to know the location of the *enzyme* within the cell, and whether the *enzyme* is present in free solution or bound to a membrane: this information may be obtained by the use of the techniques. Only then is it possible to devise a suitable *extraction* procedure.

The Extraction of Soluble Enzymes

Soluble cytoplasmic enzymes are the simplest to extract, for any disruption of the plasma membrane will enable the enzyme to pass into the surrounding medium. Soluble enzymes present in the organelles of eukaryotic cells are also easy to liberate, but the disruption of *intracellular* membranes often requires *harsher* conditions than disruption of the *plasma* membrane. To minimise the extraction of unwanted *enzymes*, cell fractionation may be carried out prior to the disruption of organelles.

The extraction procedures used depend on the type of *organism* acting as source.

Animal tissue is removed as soon after death as possible and kept cold to minimise autolysis. Obvious non-enzyme matter (e.g. fat deposits and connective tissue) is removed and the remaining tissue treated to disrupt membranes using a technique appropriate to the location of the enzyme being extracted: homogenisation in an isotonic medium disrupts the plasma membrane but leaves most organelles intact the techniques of freezing and thawing, sonication and drying in *vacuo* disrupt all plasma and intracellular membranes.

The extraction of soluble enzymes from *higher plants* varies considerably according to the type of source material. Enzymes may be extracted from cereal flours simply by placing in an aqueous medium and stirring, and from soft storage organs (e.g. the potato) by mincing and pressing out the juice through muslin or cheesecloth. Extracts from leaves, dried seeds etc. can be obtained by grinding in a mortar and then treating with a suitable medium in a pestle homogeniser. Leaves and other plant tissue may also be homogenised directly in medium by the use of a Waring blender. Alternatively, leaves may be disrupted in a Waring blender in the presence of several volumes of cold acetone (to displace water) followed by careful drying to produce an *acetone powder*: such powders may subsequently be treated with suitable medium, and the extracts will contain less chlorophyll and resinous material than would otherwise be the case. Tougher tissues, such as fibrous roots, are disrupted by grinding in a mortar with sand or ground glass. This procedure, like that of homogenisation in a Waring blender, tends to disrupt organelles.

Micro-organisms, in common with plants, have a cell wall and so tend to be more resistant to disruption than are animal cells. In general, methods of extraction fall into two main categories: drying and mechanical disruption. Drying may cause the cell to become permeable, enabling soluble enzymes to be extracted. *Air-drying* (particularly of yeast) where the cells are allowed to dry in air for several days at around 30°C, and drying in *vacuo,* where a cell suspension is placed in a vacuum desiccator for one or more days, are both accompanied by some degree of autolysis. *Lyophilisation (freeze-drying)* procedures, where a frozen cell suspension is dried by sublimation *in vacuo,* is not accompanied by autolysis, and soluble enzymes are less easily extracted from lyophilised cells than from air-dried or vacuum-dried preparations. Micro-organisms may also be dehydrated with acetone to produce powders containing easily extractable enzymes. Mechanical disruption of micro-organisms may involve grinding with carborundum beads or other abrasives, applying pressure (e.g. with Hughes or French presses), sonication, or a combination of these methods. Another procedure, which is becoming increasingly important, is to weaken the bacterial cell wall with enzymes such as lysozyme and glucuronidase, particularly where these are available in an insolubilised form and can be recovered: this treatment makes the cell more susceptible to disruption by relatively mild techniques, e.g. osmotic shock.

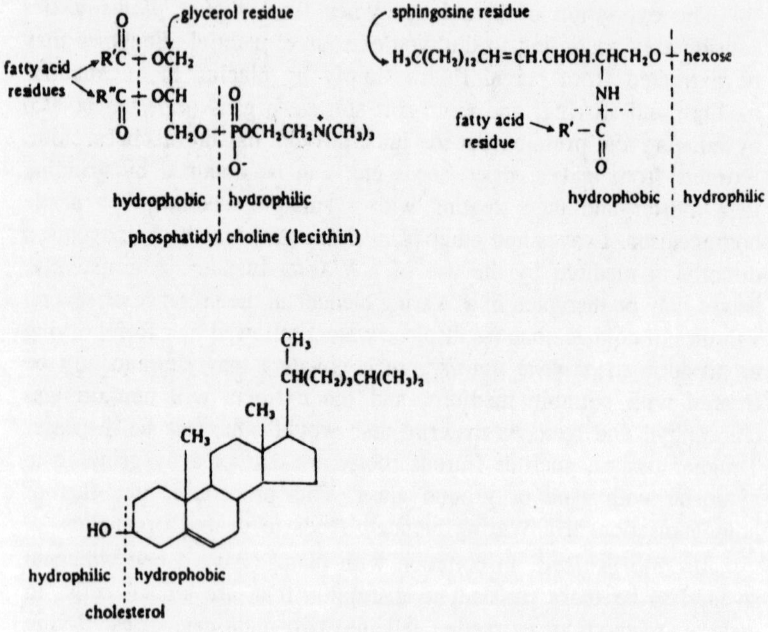

glycerol residue
fatty acid residues
$R'C(=O)-OCH_2$
$R''C(=O)-OCH$
$CH_2O-P(=O)(O^-)OCH_2CH_2\overset{+}{N}(CH_3)_3$
hydrophobic | hydrophilic
phosphatidyl choline (lecithin)

sphingosine residue
$H_3C(CH_2)_{12}CH=CH.CHOH.CHCH_2O$ — hexose
NH
fatty acid residue → $R'-C(=O)$
hydrophobic | hydrophilic

CH_3
$CH(CH_2)_3CH(CH_3)_2$
CH_3
CH_3
HO
hydrophilic | hydrophobic
cholesterol

Figure 16.1 : Some important examples of lipids found in membranes.

In general, the method of choice depends not only on the type of cell being disrupted and the sub-cellular location of the enzyme being extracted but also on the characteristics of this enzyme (e.g. its stability under the conditions of the extraction procedure): it must be admitted that the final choice is often made empirically, but a consideration of the principles discussed above can help to limit the number of possibilities.

The Extraction of Membrane-bound Enzymes

Enzymes which are bound to plasma or intracellular membranes cannot normally be extracted into the surrounding medium simply by the disruption procedures. However, some enzymes are loosely bound and some tightly bound (many actually forming an integral part of the membrane), so there can be no general rule about their extraction.

Membranes, besides containing enzymes and other proteins, are made up largely of *amphipathic lipids*, i.e. lipids which contain hydrophilic and hydrophobic regions. The main classes of membrane lipids are phospholipids (fatty acid/phosphoric acid esters of glycerol and sphingosine) and glycolipids (fatty acid esters, mainly of sphingosine, attached to a sugar). Some membranes in eukaryotic cells also contain sterols. In each case the hydrophilic portion of the molecule is small compared with the hydrophobic part (the R-groups of the fatty acid residues are long hydrocarbon chains). An amphipathic lipid may therefore be represented as a molecule with a hydrophilic head and one or more hydrophobic tails, for example:

head tail

(hydrophilic)(hydrophobic)

Although the detailed structures of membranes have yet to be established beyond doubt, the model which best fits the available evidence (including that obtained by electron microscopy and nmr spectroscopy is that of Singer and Nicolson this is termed the *fluid-mosaic* model.

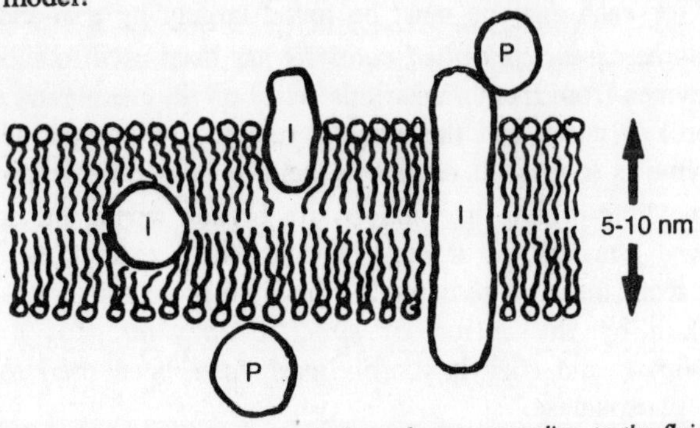

Figure 16.2 : Representation of a typical membrane, according to the fluid-mosaic model of Singer and Nicolson (P = peripheral protein and I = integral protein).

Proteins may be *peripheral* (or *extrinsic*), when they are bound to the surface of the membrane, or they may be *integral* (or *intrinsic*) when they are wholly or partly embedded in the lipid layers. Some evidence has been produced to show that the parts of polypeptide chains embedded in membranes are rich in non-polar amino acid side chains, which may form hydrophobic bonds with the lipids, whereas those parts which stick out, thus coming into contact with water, contain more polar side chains and, in some instances, are attached to hydrophilic carbohydrate units. It seems that at least some of the pores through the membrane may be lined with protein.

According to this model, lipids and proteins can move laterally about the membrane, to some extent, but find much more difficulty in moving in a transverse direction, from one membrane surface to the other.

Peripheral proteins are apparently linked by electrostatic or hydrogen bonds to the polar heads of lipids or to integral proteins and can easily be dissociated from the membrane (e.g. by treating with a solution of high ionic strength, such as 1 M NaCl, by freezing and thawing, or by sonication). Integral proteins, on the other hand, can only be extracted by breaking the hydrophobic interactions between lipid and protein (i.e. by dissociating a lipoprotein complex). Furthermore, for the extraction of an integral protein which is an enzyme, this must be done in such a way that enzyme activity is not lost during the process. A major problem is that an integral protein in its natural environment has some parts of its surface in contact with the hydrophobic membrane and other parts in contact with water, so structural changes are likely to take place no matter whether it is extracted into an organic or an aqueous medium. Again, the best method for each enzyme must be found largely by trial-and-error.

In some cases, *controlled autolysis* has been used to successfully free enzymes from membranes; this relies on the membrane structure being broken down and the enzyme released before the activity of the enzyme is lost, so it cannot be held to be a reliable procedure.

Some lipid-protein interactions are broken during the formation of *acetone powders*, so it may be possible to extract the required enzyme from these. Alternatively, the lipid-protein interactions may be broken by the action of *specific enzymes*: for example, phospholipase and triacylglycerol lipase have been used to extract alkaline phosphatase.

Of more general significance, *detergents* have been used to disrupt membranes and extract lipoprotein fragments into aqueous media as components of micelles (aggregates of amphipathic molecules arranged with their hydrophobic parts in the interior of the micelle and their hydrophilic parts on the surface). These detergents include natural bile salts such as cholate; non-ionic synthetic detergents like Triton X-100; and ionic synthetic detergents which may be zwitterionic, as in the case of CHAPS, acidic like sodium dodecyl sulphate (SDS) or basic, e.g. CTAB. In addition to the protein-detergent micelles, molecules of a detergent will themselves form micelles if present at a concentration above the critical micelle concentration (CMC). In general, the bile salts form the smallest micelles (molecular weight below 2000) and have the highest CMC (1-10 mM) while the non-ionic detergents form the largest micelles (molecular weight up to 100 000) and have the lowest CMC (in the order of 0.1 mM). A knowledge of these characteristics may be helpful in the further purification of the extracted enzyme. The choice of detergent will be influenced by whether or not the enzyme is required in an active form, and what further purification procedures are intended. A detergent of particular importance in the investigation of membranes is SDS, which will sever most protein-lipid (and protein-protein) interactions and enable the molecular weights of the liberated proteins to be determined. However, for extraction of an active enzyme, non-ionic detergents are usually preferred.

Lipoprotein fragments may also be extracted by the use of *organic solvents*, and some free proteins may be liberated on subsequent dispersion in aqueous media: for example, xanthine oxidase has been obtained from lyophilised milk protein by a technique involving ether extraction. Such procedures might require evaporation of the organic medium and dissolving the enzyme in a detergent solution.

The solvent most widely used for the extraction of enzymes from membranes has been *n-butanol*. This has both lipophilic and hydrophilic character, so can act like a detergent in removing lipoproteins from membranes and, further, tends to dissociate the protein and lipid components. Most membrane proteins are found in the aqueous phase of a butanol-water two-phase system, with the lipids in the butanol phase. Different groups of enzymes may be separated from membrane lipids by treatment with n-butanol under different conditions: for example, alkaline phosphatase has been extracted at pH 5-6, γ-glutamyltransferase at pH 7 and urate oxidase at pH 10. In general, however, organic solvents tend to cause some degree of protein denaturation, so are not ideal for the extraction of active enzymes.

It will be realised that enzymes can exist in cells complexed to nucleic acids or carbohydrates, but, unless they are also linked to lipids in membranes, their extraction will be relatively straightforward.

The Nature of the Extraction Medium

Once cells have been disrupted, by whatever method, the contents are extracted with several volumes of aqueous medium. In the case of enzymes which are an integral part of membranes the original extraction may be in organic solvent but the liberated enzyme must then be further extracted into aqueous medium.

In general, a considerable amount of debris (mainly membrane fragments) is likely to be present, so it is essential that the enzyme of interest can be easily separated from this by being soluble in the extraction medium.

Four main factors govern the solubility of proteins: salt concentration, pH, the organic solvent content and temperature. The temperature of the medium is usually kept below 4°C, despite the reductions in solubility that this entails, in order to minimise loss of activity of the enzyme.

Proteins are least soluble at their *isoelectric point*, so the extraction of an enzyme must be carried out at a pH value far from its isoelectric point, but nevertheless at a value where the enzyme is stable. The salt and organic contents of the medium are chosen, largely on an empirical basis, to ensure that the enzyme is soluble but many other proteins are not, thus achieving a preliminary purification. For example, Cohn and co-workers showed that glutamate dehydrogenase, arginase, alkaline phosphatase and acid phosphatase from a bovine liver homogenate were soluble in 19% ethanol at pH 5.8 and ionic strength 0.02, but catalase, peroxidase and D-amino-acid oxidase were not; however, these three enzymes were soluble at pH 5.8 in the absence of ethanol and at ionic strength 0.15; also, various proteinases which were insoluble in both of these media were soluble if the pH of the second medium was increased to 7.4.

The solubility of an enzyme can sometimes be increased by the addition of its *substrate*: for example, alkaline phosphatase is more soluble in the presence of β-glycerophosphate than otherwise. Hence the extraction of an enzyme can be facilitated by adding its substrate to the medium.

Stabilisers, such as *dithiothreitol (Cleland's reagent)* or *mercapto-ethanol* are sometimes added to extraction media to prevent oxidation of suiphydryl groups and thus loss of enzyme activity. *Inhibitors* of proteolytic enzymes (e.g. DFP) may also be added to prevent these attacking the enzyme of interest. Polyols (e,g. glycerol) or chelating agents (e.g. EDTA) also have a use in stabilizing solubilised enzymes.

PURIFICATION OF ENZYMES

Preliminary Purification Procedures

As mentioned in a preliminary purification step is usually included in the procedure used to extract an enzyme. For example, some degree of cell fractionation is often carried out before extraction, so that only enzymes in the same sub-cellular compartment as the relevant enzyme are extracted. Also, the composition of the extraction medium is chosen so that the enzyme of interest is soluble but many others are not.

Extracted nucleic acids may be precipitated by treatment with basic substances such as streptomycin or protamine, or with $MnCl_2$ or $MgCl_2$.

All precipitates and cell debris are then removed by centrifugation and discarded. Polysaccharides may also be removed by high-speed centrifugation, although the results are not always satisfactory. An alternative way to dispose of polysaccharides is to break them down into small units with enzymes such as a-amylase; nucleic acids may similarly be broken down by the use of nuclease enzymes.

The next stage of purification is usually to precipitate the enzyme of interest from solution, thus separating it from mono- and oligo-saccharides, nucleotides, free amino acids etc. and from many other proteins which remain in solution. This may be achieved by altering the pH and the organic or salt concentrations of the medium. For example, the pH may be adjusted to the isoelectric point of the enzyme. Alternatively a high salt concentration can be introduced: ammonium sulphate is often used for this purpose, because it is extremely soluble in water. The salt concentration is usually increased in stages, the aim being to precipitate other proteins (which are then discarded) before precipitating the fraction containing the relevant enzyme. This is then redissolved in water, and residual ammonium sulphate removed by dialysis.

Although a considerable degree of purification can be achieved by these procedures, many other proteins will still be present because of overlap of solubility ranges. One possible way to remove some of these is to raise the temperature of the medium for a few minutes to a value where the enzymes being purified is known to be stable, but others might be denatured and precipitate from solution. Enzymes are often particularly stable in the presence of their substrates, so the relevant substrate can be added to increase the effectiveness of this procedure.

Further Purification Procedures

The crude extract, partially purified may be treated in a variety of ways to increase the purification of the relevant enzyme.

Adsorbent gels (e.g. zinc hydroxide) have been used to remove pigments from enzyme preparations. A mixture of enzymes may also be adsorbed by a suitable gel (e.g. aluminium hydroxide) and then fractionated by elution with buffers of increasing ionic strength. *Partition chromatography* has similarly been used to separate mixtures of enzymes: for example, ribonuclease was purified by Martin and Porter on columns packed with kieselguhr, using a two-phase system consisting of ammonium sulphate, ethyl cellosolve and water; the sample was applied to a column which had been pre-equilibrated with the organic phase, and then eluted by the aqueous phase, separation of components being on the basis of relative solubility in the two phases.

Such techniques played an important part in the development of enzymology, but have now been superseded by others.

Electrophoresis is mainly an analytical procedure, since it is ideally suited to the separation of small amounts of material, but it has also been used for the purification of proteins. The rate and direction of migration of a protein in an electric field depends on its net-charge at the pH used and also on the size of the molecule, since this imposes some resistance to movement. In *zone electrophoresis*, a mixture of proteins is introduced at a common point (the *origin*) and allowed to move in an electric field, usually in a horizontal direction, each molecule travelling in the same zone as others of the same charge and size. To minimise diffusion the whole process is usually carried out in a solid but porous *support medium* (e.g. starch gel) through which the buffer and proteins permeate. When separation has been achieved, as indicated by cutting a narrow longitudinal section from the support medium and staining for proteins (e.g. with Amido Black) then the rest of the support medium may be cut into strips across the direction of travel and the contents of each extracted by elution. Alternatively, zone electrophoresis may be carried out in a vertical direction in a closed column with, for example, powdered cellulose as the support medium; after electrophoresis has been completed, a tap at the bottom of the column is opened to enable the liquid contents to be run out and collected in fractions.

Some forms of electrophoresis capable of giving extremely high resolution, e.g. *isotachophoresis and isoelectric focusing*, which are employed at present mainly for analytical purposes, are likely to find increasing use in preparative work.

Cation exchange chromatography on columns packed with carboxylated polystyrene has been used in the successful purification of stable, low molecular weight basic proteins such as lysozyme and ribonuclease. However, *cellulosic ion exchange chromatography* has proved of more general application to the separation of proteins. Such resins, in contrast to most other ion exchange resins, are relatively hydrophilic and have an open structure readily penetrated by large molecules such as prcteins. The most frequently used are diethylaminoethyl (DEAE)cellulose which is an anion exchanger, and carboxymethyl (CM) cellulose, a cation exchanger. The DEAE group, -$OC_2H_5NH(C_2H_5)_2$, is highly positively charged at pH 6-8, so DEAE-cellulose is most useful for the chromatography of proteins which are negatively charged in this range; similarly, CM-cellulose, cellulose-OCH_2CO2, is most applicable for the separation of proteins which are positively charged at around pH 4.5.

Elution of proteins from the columns may be brought about by changes in either salt concentration or pH: as the concentration of salt (e.g. NaCI) increases, protein is displaced from DEAE-cellulose by the anion (Cl^-) and from CM-cellulose by the cation (Na); if the pH is altered over the relatively narrow working range, proteins are eluted as their isoelectric point is reached, since they then have no net charge with which to bind to the resin.

Ion exchange chromatography has been used in the purification of membrane proteins in non-ionic or zwitterionic detergents as well as for the purification of soluble enzymes.

The buffering capacity of ion exchangers may be exploited to create a linear pH gradient in the column and separate proteins by *chromatofocusing*. This technique has features in common with isoelectric focusing.

Gel filtration (size-exclusion or *molecular-sieve chromatography*), another important technique in enzyme purification, is carried out with columns packed with swollen gels which separate components of a sample on the basis of molecular size: molecules too large to enter the pores of the gels will pass quickly through the columns; smaller molecules will pass through the column more slowly, the actual speed for each component being dependent on the ease with which its molecules can pass into the gels and be thus retarded. Gels commonly used include cross-linked dextrans (Sephadex), cross-linked agarose

(Sepharose) and cross-linked polyacrylamide (Biogel). These are graded according to pore size, and thus to the size (and hence molecular weight) of protein molecule which can enter. For a successful separation, the enzyme of interest must not be too large to penetrate the gel, because it would then pass straight through the column together with all other proteins of similar and greater molecular size.

Eluting buffers should be of high ionic strength (e.g. 20 mM) to counteract the few charges which may be present on the gel: apart from that, the only criterion for the buffer is that the proteins are stable in it: no pH or salt gradient is employed, since the separation is by size alone. However, it should be noted that some of these gels are available with attached ion exchange groups (an example is DEAE-Sephadex) in which case the principles of elutions are as for cellulosic ion exchange chromatography.

In the case of membrane proteins solubilised in detergents, gel filtration is widely used to separate protein-detergent micelles from excess detergent: this is particularly easy to achieve if detergent molecules are kept in the monomeric form, e.g. by ensuring that the detergent concentration is less than the CMC, or by adding a small amount of a bile salt.

Affinity chromatography is a bio-specific process and thus ideally suited for the separation of one protein from all others, including those which resemble it so closely in physical characteristics that separation by any other procedure is extremely difficult. The column is packed with an inert matrix (e.g. agarose) to which ligands for the required enzyme have been attached. These immobilised ligands may be, for example, substrate analogues, and must be covalently linked to the matrix in such a way that they can still bind to their enzyme, if this is present: this usually means that a *spacer arm* (e.g. a hydrocarbon chain) must be included between matrix and ligand, joining to the ligand at a point which does not interact with the enzyme, if steric hindrance to the formation of an enzyme-ligand complex is to be avoided. The attachment of the spacer arms to -OH groups of the matrix is often achieved by the use of CNBr reagent.

The protein mixture is applied to the column, and the relevant enzyme is trapped by the immobilised ligands while all other proteins pass through and are discarded. The enzyme is then liberated from the column either by eluting with a *deforming buffer* at a pH which changes the characteristics of the enzyme and no longer allows it to bind to the immobilised ligand, or by the use of a *competitive counter-*

ligand, which displaces the immobilised ligand on the enzyme. In both cases, the enzyme passes through the column and can be collected, now free of other proteins. An example of an enzyme extracted by the use of such a technique is fl-galactosidase with the substrate-analogue p-aminophenyl β-D-galactoside as ligand, the deforming buffer being 0.1 M borate at pH 10.

A closely related procedure is *immunoaffinity chromatography*, where the immobilised ligand is an antibody specific for the enzyme which is being purified. The use of an immobilised monoclonal antibody as ligand gives even greater specificity than use of a more conventional antibody.

The bio-specific nature of these techniques means that a high degree of purification can be achieved; on the other hand, it also means that a polymer-ligand complex prepared for the purification of one enzyme is limited to that particular application.

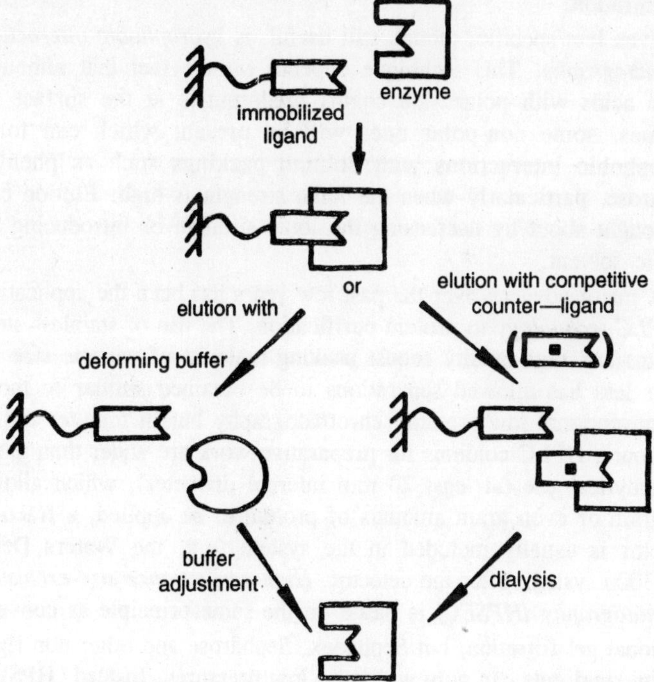

Figure 16.3 : Diagrammatic representation of the processes involved in affinity chromatography; the treatment with deforming buffer and with competitive counter-ligand are alternatives.

Another rapidly developing form of affinity chromatography involves the use of reactive ***triazine-based dyes***, e.g. Cibacron Blue F3G-A and dyes of the Procion series, as ligand. Under alkaline conditions the chlorotriazine dye is linked, usually directly, to a matrix such as agarose by a triazine bond; the reaction involves a hydroxyl group on the matrix and a chloride on the dye. Such immobilised dyes bind a wide range of NAD- and NADP-dependent dehydrogenases and other enzymes. They are, of course, less specific than biological group-specific ligands, and their specificity is difficult to predict, but they have many advantages: they have a greater protein-binding capacity, they are extremely resistant to chemical and enzymic degradation, they can be used and re-used in a variety of applications, and the triazine bond is more stable than the isouronium linkage introduced by the CNBr activation of agarose. Elution of enzyme from the ligand is usually achieved by changing pH or increasing salt concentration.

Even less specific, though still useful, is ***hydrophobic interaction chromatography***. This technique depends on the fact that although amino acids with polar side chains predominate at the surface of enzymes, some non-polar ones will be present which can form hydrophobic interactions with column packings such as phenyl-Sepharose, particularly when the ionic strength is high. Elution can be brought about by decreasing the ionic strength or introducing an organic solvent.

A major advance over the past few years has been the application of *HPLC technology* to protein purification. The use of stainless steel columns and high-quality robust packing material of particle size 10 um or less has allowed separations to be obtained similar to those for conventional low-pressure chromatography but in minutes rather than hours. HPLC columns for preparative work are wider than those for analytical use (at least 20 mm internal diameter), which allows milligram or even gram amounts of protein to be applied; a fraction collector is usually included in the system (e.g. the Waters Delta Prep 3000 system) after the detector. *High-performance size-exclusion chromatography (HPSEC)* is based on the same principle as conventional gel filtration, but Sephadex, Sepharose and other non-rigid or semi-rigid gels can only withstand low pressures. Instead, HPSEC often utilises rigid spherical beads of porous silica with bonded hydrophilic polar groups. *High-performance ion exchange chromatography (HPIEC)* utilises amines as anion exchangers and sulphonic or carboxylic acids as cation exchangers, each bonded to some rigid support such as silica. ***High-performance liquid affinity chromatography*** utilises ligands bound to supports such as epoxy-silica:

$$\underset{\text{silica}}{R\text{—}OH} + \underset{\text{epoxy-silane}}{(CH_3O)_3Si\text{—}(CH_2)_3OCH_2\overset{O}{\overbrace{CH\text{—}CH_2}}} \longrightarrow \underset{\text{epoxy-silica}}{R\text{—}O\text{—}\overset{|}{\underset{|}{Si}}\text{—}(CH_2)_3OCH_2\overset{O}{\overbrace{CH\text{—}CH_2}}}$$

$$\xrightarrow{\text{ligand—}NH_2} R\text{—}O\text{—}\overset{|}{\underset{|}{Si}}\text{—}(CH_2)_3OCH_2\overset{OH}{\overset{|}{C}}H.CH_2NH\text{—ligand}$$

Proteins may also be separated by ***reverse-phase HPLC*** on alkylsilica columns, the eluting solvents being buffered aqueous and organic mixtures.

When after the use of one or more of these procedures it is considered that the relevant enzyme has been separated completely from other proteins mineral salts and any small molecules present may be removed by *dialysis*, if this is desired.

The purified enzyme preparation is likely to be quite dilute, so it might require concentrating before being used for the purpose for which it was prepared. One convenient method for doing this is *lyophilisation*, followed by the redissolving of the enzyme in a small volume of liquid. Another suitable procedure is *ultrafiltration*: this involves forcing the solvent molecules through a membrane of chosen porosity (e.g. by the application of nitrogen gas pressure) and thus separating them from the protein molecules, these being too large to pass through the pores.

The ideal way to complete a purification procedure is to *crystallise* the enzyme, if this is possible. Sometimes it may be achieved by adding not quite enough ammonium sulphate to cause precipitation of the enzyme and leaving this solution in a coldroom for several days. In general, successful conditions have to be determined empirically by attempting crystallisation at a variety of salt and organic solvent concentrations and pH values, and by showing much patience.

Membrane enzymes may be inactive after purification unless re-introduced into a phospholipid environment. For example, phospholipid may be added to purified protein-detergent micelles and concentrations adjusted so that the phospholipid replaces detergent in the micelles;

released detergent, in the monomeric form, may then be removed by gel filtration, dialysis or ultrafiltration. An example of a membrane enzyme that has been characterised is Na^+-K^+ATPase.

Criteria of Purity

At each stage of a purification procedure, an assay for the enzyme being purified should be performed on all fractions and its *specific activity* in each fraction determined. This will show, irrespective of preconceived ideas, precisely which fractions contain the important enzyme and will enable the degree of purification to be calculated. The *total* activity of this enzyme will be unchanged as a result of the purification step, unless some has been lost during purification; under no circumstances can it be increased. However, some of the contaminating proteins which were originally present in the same preparation as the enzyme should now be in different fractions. Hence, if the fractions containing the relevant enzyme are combined, the total activity of the enzyme in the combined fraction should be the same as that started with, whereas the total amount of protein in this fraction will be less than that originally present. The *specific* activity of the enzyme in the combined fraction should therefore be greater than that in the preparation before purification, and the increase in specific activity will be a measure of the purification achieved.

With each successive purification step, the specific activity of the fraction (or fractions) containing the enzyme should be greater than before until complete purification is achieved and the specific activity reaches a limiting value.

However, the finding of the same specific activity value before and after a purification step does not necessarily mean that the enzyme preparation is completely pure: it could simply mean that contaminating proteins have passed through the procedure in the same fraction as the enzyme. Similarly, crystallisation cannot be taken as proof that only one protein is present, for many mixed protein crystals have been found. Hence other criteria of purity have to be considered.

If a particular enzyme is known to be a possible contaminant, then a logical step is to carry out an assay for that enzyme and demonstrate its absence in that way. However, that would not exclude the possibility that other contaminants were present.

Analytical ultracentrifugation is widely used to investigate the purity of enzyme preparations. With high centrifugal fields (e.g. 500 000 g), diffusion effects are largely overcome hence, as a randomly distributed group of identical protein molecules sediment in an *analytical cell* under the effect of such a field, a sharp boundary is formed between the portion of the medium which has been cleared of protein molecules and that which still contains them. This boundary refracts light, and its position can be determined through a quartz window in the cell: for example, it shows up as a peak if the refractive index gradient at each point in the cell is measured using the *Schlieren optical system*. The movement of this boundary as sedimentation proceeds can be constantly monitored, and automatically recorded, and the data used to calculate the characteristics of the molecule in question. If the protein is pure, only one boundary should be observed: detection of extra boundaries suggests the presence of contaminants.

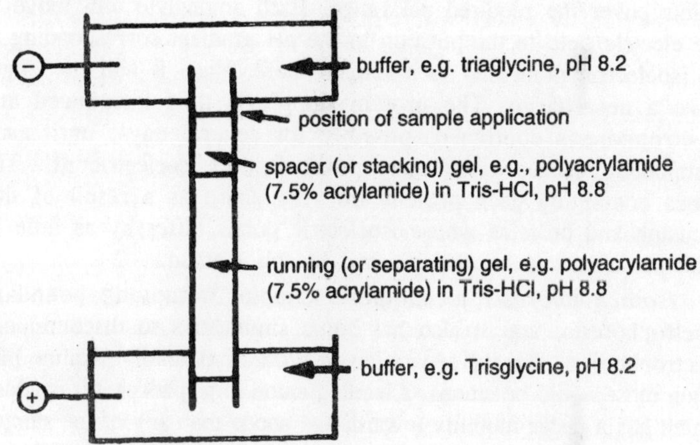

Figure 16.4 : Diagrammatic representation of an apparatus set up for discontinuous (disc) electrophoresis of proteins.

Electrophoresis in various forms is very effective in separating mixtures of proteins so it is used extensively to investigate the purity of enzyme preparations. Of particular importance in this context is *discontinuous (disc) electrophoresis*, where the electrodes are placed in separate buffer reservoirs linked vertically by gel (usually polyacrylamide) supported by glass plates or cylinders. The discontinuity is in the gels and buffers used, pH differences being used to concentrate the sample into a very narrow band as it passes through a large-pore *spacer gel* on its way to the *running gel* where the actual separation of the sample components takes place. This running gel has smaller pores than the spacer gel, and separation is achieved by a combination of electrophoresis and gel filtration. On completion of the electrophoresis and after staining for proteins, it is easy to see if more than one band is present, for each is extremely narrow and well-defined.

Capillary electrophoresis, in which charged components are separated in media inside small capillaries, is becoming increasingly important. This provides a rapid and efficient analytical technique.

Another relevant technique is *isoelectric focusing*, which is an example of *moving boundary electrophoresis* rather than zone electrophoresis. A pH gradient is set up between the electrodes by allowing an acid (e.g. phosphoric acid) to diffuse in from the anode end and a base (e.g. ethanolamine) from the cathode end. This can be stabilised by introducing a mixture of *ampholytes* (e.g. synthetic and natural amino acids) selected so that their individual isoelectric points cover the required pH range. Each ampholyte will move in the electric field to the position in the pH gradient corresponding to its isoelectric point and then remain there, since it will no longer have a net charge. The protein sample is then introduced and electrophoresis continued, possibly for several days, until each component reaches a stationary position at its isoelectric pH. The zones containing each protein are very sharp as a result of this focusing and proteins whose isoelectric points differ by as little as 0.02 pH units can be distinguished by this method.

Isotachophoresis is another example of moving boundary electrophoresis, and it also has some similarities to discontinuous electrophoresis. For the separation of proteins at mildly alkaline pH, when most would be anions, a leading anion (e.g. phosphate) is added which has a faster mobility towards the anode than any of the sample anions. A trailing or terminating anion (e.g. glycine), with a slower mobility than any in the sample, is also added. The system is buffered by a common-counterion, e.g. Tris. The leading and terminating anions are applied at different sides of the sample, the leading anion nearest the anode. A high voltage is applied and, when steady-state is achieved, all components will migrate towards the anode in discrete zones at the same velocity ('iso tacho' being Greek for 'same speed'), the sample anions arranged in order of their mobilities. Spacers similar to the ampholytes used in isoelectric focusing may be added to separate the zones of sample anions. Isotachophoresis may be carried out in a purely aqueous environment in a capillary tube or, particularly for preparative work, in columns packed with gel (e.g. polyacrylamide) of pore size large enough to minimise molecular sieving.

Analytical HPLC procedures may also be used to investigate the purity of a protein preparation.

No single technique, however sensitive, can establish enzyme purity because of the possibility that a contaminating protein might behave in an identical fashion to the important protein under the conditions used. However, if several characteristics are investigated, and the conditions of investigation varied, the chances of the contaminant avoiding detection are reduced. Therefore, evidence of purity must be obtained by the use of several of the above procedures before it can be concluded that the sample is indeed pure.

DETERMINATION OF MOLECULAR WEIGHTS OF ENZYMES

When an enzyme has been purified, its molecular weight may be determined.

Gel filtration, which separates molecules on the basis of size, provides one way of doing this. A packed column is calibrated by applying proteins of known molecular weight to the top of the column and determining the volume of buffer required for the elution of each protein: as each protein leaves the column there should be an absorbance peak at 280 nm in the column eluate, thus providing a simple way of monitoring the elution. From the data obtained, a graph may be drawn of elution volume against molecular weight. The protein of unknown molecular weight is then passed through the column and its elution volume determined exactly as for the marker proteins. Hence, by reference to the calibration graph, the molecular weight of the protein may be estimated. One source of error in this procedure is that it takes no account of molecular shape: in fact, a protein molecule which is somewhat elongated will pass down the column more slowly than a spherical one. of the same molecular weight.

Ultracentrifugation is also widely used for the determination of molecular weights of proteins. For a spherical protein of molecular weight M_r sedimenting through a medium in a centrifugal field, the centrifugal force per grammole is M_r (1 - *ap)co*^{2}*r* where v is the partial specific volume of the molecule (i.e. the increase in volume in cm^1 when 1 g solute is added to a large volume of solvent; the value for a protein is usually about 0.74). The centrifugal force is opposed by an equal frictional force per gram-mole of sedimentation rate where R is the gas constant (in ergs per mole per degree), T is the absolute temperature and D is the diffusion constant of the molecule.

This is known as the *Svedberg equation*, after the pioneer of ultracentrifugation.

The rate of movement of the boundary separating the protein-free zone from the rest of the medium can be determined, and this is a measure of the sedimentation rate. Hence the sedimentation coefficient (s) can be calculated, since co^2 and r are easily determined: for most proteins, s lies in the range 1-200 S. The diffusion constant *(D)* can similarly be calculated from moving boundary determinations, this time in the absence of a centrifugal field.

Once these and the other relevant factors have been determined, the molecular weight of the protein can be calculated by the use of the Svedberg equation. However, as with gel filtration, there is a possible source of error in the assumption that all protein molecules are spherical.

A technique which is commonly used to determine molecular weights of proteins, and which does not assume that the molecules are spherical is *SDS-gel electrophoresis*. SDS (sodium dodecyl sulphate, $CH_3(CH_2)_{11}$ SO_3 Na^+) is an anionic detergent whose hydrocarbon chain can become linked to hydrophobic regions in the interior of globular proteins, leaving the ionised sulphonate ($-SO_3$) group jutting out into the surrounding medium. Many SDS molecules can bind in this way to a single protein molecule, disrupting its natural shape and giving it a large negative charge, regardless of the net charge originally possessed by the protein. Large protein molecules will complex with more SDS than small ones, so all will have roughly the same charge/mass ratio, and all will have much the same shape. Therefore, if molecules of different proteins are complexed with SDS and subjected to zone electrophoresis in an unrestrictive medium, all should travel together, regardless of their original characteristics. However, if electrophoresis is performed in a support medium with a restrictive pore size (e.g. in polyacrylamide gel), the complexes from each of the different proteins will be separated, entirely on the basis of size. Since the size of each protein-SDS complex must be dependent on the original size of the protein molecule, a system can be calibrated by the use of marker proteins of known molecular weight; hence the molecular weight of a test protein may be determined. One point that should be borne in mind, however, is that SDS tends to break all weak polypeptidepolypeptide interactions, so, for an oligomeric protein, this technique may well give the molecular weight of the sub-unit rather than of the complete protein.

10

ENZYMOLOGY, NONAQUEOUS

Nonaqueous enzymology is concerned with the utilization and understanding of enzymes in essentially organic environments. The development of structure-function-environment relationships is central to our understanding of how enzymes work as catalysts in vivo and in vitro. This article introduces the overall subject, and presents briefly some of the advances made to date.

Living organisms synthesize a wide array of enzymes, which catalyze a myriad of reactions both inside and outside the cell. The environment of the enzymes not associated with a membrane is essentially aqueous. For this reason, conventional biocatalysis is carried out in aqueous media, and it is not surprising that most methods developed to study enzyme performance are water based. Recently, however, interest has been focused on using enzymes to catalyze reactions in organic media. Industrial enzymatic processes offer the advantage of mild reaction conditions, extreme selectivity, and environmental acceptability. If an enzyme could function in an essentially organic environment, increased ease of product recovery, increased hydrophobic reactant solubility, and reduced microbial contamination would be properties contributing to wide applicability. This entry focuses on another remarkable feature of enzyme catalysis in solvents other than water: namely, that altering the solvent can have a dramatic effect on enzyme function, enabling the production of catalysts by design via "solvent engineering."

Modern-day nonaqueous enzymology evolved in a number of

distinct phases. Initially, water-miscible organic solvents, such as acetone and ethanol, were added to aqueous enzyme solutions to determine the maximal concentrations of solvents the enzymes could tolerate (-50% for most conventional enzymes). Next, biphasic mixtures in which an aqueous solution of enzyme was emulsified in such water-immiscible solvents as chloroform and ethyl acetate were evaluated. This approach was further developed by the use of reversed micelles to stabilize the enzyme in such a mixture. In 1966 Dastoli and Price demonstrated that freeze-dried enzyme powders retained significant activity when suspended in a wide variety of organic solvents. In the early 1980s, Klibanov revolution-ized the field of nonaqueous enzymology by demonstrating that enzymes could catalyze novel reactions in organic media, and even exhibit enhanced thermostability. Klibanov's work led to the development of a new approach for enzyme studies: not only can we investigate how an enzyme works, but we can also determine whether it can be fine-tuned by altering its environment.

Catalysis is dependent on the intact three-dimensional structure of a protein. Water plays a critical role in all noncovalent interactions that maintain an enzyme in its catalytically active form. It was assumed therefore that replacement of water by a nonaqueous medium would result in the loss of enzyme activity. However, an enzyme freeze-dried (lyophilized) from aqueous medium retains a small amount of water (—15% by weight, and less than a monolayer per molecule of protein). Thus a dry enzyme powder, suspended in a nonaqueous dispersant, can be protected from bulk solvent by a shell of water, enabling the enzyme to function. Because of the water associated with enzymes suspended in anhydrous solvents, the systems are often termed "essentially anhydrous."

Lyophilized enzyme powders suspended in organic solvents have been used to perform a variety of important catalytic functions such as esterifications, transesterifications, aminolyses, oximolyses, racemate resolutions, oxidations, reductions, peptide syntheses, polymerizations, and depolymerizations. The insolubility of enzymes in organic solvents, resulting in a biphasic system, also results in such advantages as ease of enzyme recovery by filtration. The absence of water leads to improved enzyme thermostability (water is a reactant in many processes that irreversibly denature enzymes), and the reduction of undesirable side reactions that require water as a substrate. Since heterogeneous systems can be diffusionally limited, mixing and sonication are usually employed. A recent development in nonaqueous enzymology is the use of polyethylene glycol (PEG) modified enzymes, which are soluble in organic solvents. These enzymes form homogeneous reaction systems and thereby eliminate the problems of diffusion. Figure 1 summarizes the various approaches that can be utilized to enable enzymes to function in organic solvents. While the given methods involve the general principles of using enzymes in connection with an organic solvent or compressed gas (supercritical fluid), each system offers several advantages and disadvantages that dictate the choice of system.

SOLVENT ENGINEERING VERUS PROTEIN ENGINEERING

The most remarkable feature about the effect of a solvent on an enzyme is that the environment can be used to alter the "natural" activity and specificity of the catalyst. This approach to catalyst design has been termed "solvent engineering." Both solvent and protein engineering (where one changes the catalyst itself) have the same ultimate aim, that is, the manipulation of an enzyme's functional properties. Table 1 lists the goals of solvent and protein engineering, many of which have been achieved using solvent engineering, although the field is still in its infancy. In this discussion we focus on the use of dry enzyme powders in organic solvents.

Table 10.1 : Typical Goals of Solvent and Protein Engineering

Goal	Success Using Solvent Engineering	Success Using Protein Engineering
Modification of catalytic	X	X activity
Modification of enzyme	X	X selectivity
Control of enzyme	X	X enantioselectivity
Increase of enzyme	X	X thermostability
Modification of enzyme pH	X	X dependence
Control of allosteric function	X	
Utilization of enzymes in non-	X	X native reactions
Stabilization of enzymes	X	against denaturants

Effect of Solvent on the Physical Properties of An Enzyme

A complete characterization of enzyme function requires a detailed study of the effect of environment on protein structure, activity, and specificity. Morphological studies on protein powders suspended in organic solvents have shown considerable changes in the powder as a function of the water associated with the preparation. In general, most protein powders are reported to be flaky or spherical, with an average diameter of 100 μm. Structural studies on proteins suspended in organic solvents have been performed using electron paramagnetic resonance spectroscopy, nuclear magnetic resonance spectroscopy, infrared spectroscopy, and circular dichroism. All studies conclude that the structural integrity of enzymes is not altered in the presence of a hydrophobic bulk solvent. Organic solvents can, however, affect enzyme structure and function by directly interacting with water molecules bound to the enzyme. Hydrophilic solvents can easily "strip" the "essential" water from the surface of enzyme molecules, thus removing the protective layer from the enzyme and enabling direct interaction between nonaqueous solvent and protein. These direct interactions may cause fluctuations in enzyme structure, inactivating the protein.

The effect of solvent on the mobility of proteins has also been investigated, with interesting results. Proteins in water are like a "Swiss cheese"; the holes are full of water, which lubricates the protein. Proteins are by their nature very flexible, and the activity of the molecule is in large part determined by the flexibility of the molecule. An enzyme suspended in a hydrophobic solvent will usually retain enough water, and thus flexibility, to maintain function. In hydrophilic solvents, however, the loss of water from the powder to bulk solvent results in rigidity and decreased activity. This limitation can be overcome to some degree by simply adding extra water to the solvent.

Engineering Enzyme Activity and Substrate Specificity

Electrostatic forces such as ion pairing, hydrogen bonding, and hydrophobic and electrostatic interactions influence substrate binding and catalysis. Conventional protein engineering uses such meth-ods as site-specific mutagenesis to add or remove residues at or near the active site, thus controlling enzyme specificity and stereoselectivity. Solvent engineering allows control of all these properties by controlling the reaction medium.

The nature of the reaction medium affects enzyme activity by influencing kinetic parameters such as catalytic turnover, enzymesubstrate binding, and the overall efficiency of the enzyme. Upon transition from aqueous to nonaqueous media, turnover decreases and binding increases. Substrate and solvent hydrophobicity often determine the extent of the effect of the medium on enzyme efficiency. As substrate hydrophobicity increases, the substrate will have a greater tendency to partition into the solvent rather than into the now relatively hydrophilic enzyme active site, and more substrate will be required to saturate the enzyme.

It has also been suggested that solvent can modulate specificity. Enzymes like chymotrypsin and subtilisin, whose substrate-binding pockets are lined by hydrophobic amino acid residues, show a preference for hydrophobic nonpolar substrates in water. However, in organic media the same enzymes show a preference for hydrophilic substrates. This is because polar substrates can easily partition into the enzymes' active sites, which are less hydrophobic than the reaction medium. Hence, enzymes in organic media often show altered substrate specificity compared to that in water.

In general, enzymes retain their stereoselectivity in organic media. It is hypothesized that L-isomers bind "correctly" and nisomers "incorrectly" to the active site. The consequent release of water from the active site, due to the binding of L-isomers, is therefore, an energetically favored process in water rather than in organic solvents. As a result, the transition from aqueous to nonaqueous medium depresses the catalytic efficiency of L-isomers more than it does that of n-isomers. Hence, in some cases a nonaqueous medium "relaxes" an enzyme's enantioselectivity. In addition, solvents can be used to fine-tune the enantioselectivity to a predetermined degree.

Two novel features of enzymes in nonaqueous media are inhibitor-induced enzyme memory and pH memory. When an enzyme is lyophilized in the presence of a competitive inhibitor, the high rigidity of enzymes in organic media allows the enzyme to maintain an activated form even after the ligand has been removed. The phenomenon of "pH memory" refers to the ability of charged groups on the enzyme to retain the ionization state corresponding to the pH of the aqueous solution from which it was lyophilized. These two phenomena show how enzyme pretreatment offers a means to regulate enzyme activity and substrate specificity in nonaqueous environments.

Engineering Protein stability

A relatively common approach in the design of improved biocatalysts is to alter the protein itself, using either chemical modification or genetic engineering. Improvements in stability toward denaturants, oxidants, proteolysis, and high temperatures are frequently the desired end results of such strategies. Interestingly, enzymes in organic solvents often show enhanced thermostability (enzymes are stable at temperatures of up to 150°C when suspended in anhydrous organic solvents under optimized conditions). Irreversible protein denaturation is often limited by the rates of disulfide bond interchange, deamidation of glutamine and asparagine residues, and hydrolysis of peptide bonds. Each of these reactions requires water as a substrate. The absence of water in nonaqueous media, therefore, positively affects enzyme stability.

PROTEIN ENGINEERING FOR NONAQUESOUS MEDIA

A recent trend in nonaqueous enzymology has been the use of protein engineering techniques to design enzymes with improved stability and activity in organic media. Rules for redesigning proteins are formulated via a deep understanding of protein-solvent interactions, as well as the effect of the solvent on the noncovalent forces that affect protein structure and function. Efforts at redesigning proteins have been directed at improving the conformational stability and the compatibility of the protein surface with the organic media. It has been suggested that the former may be accomplished by the introduction of residues that can form new disulfide bridges, increased van der Waals and electrostatic interactions, and improved hydrogen bond formation. Replacement of both charged surface amino acids, which require water for solvation, and surface hydrogen bonding sites, which cannot be satisfied by nonaqueous media, have been demonstrated to improve proteinsolvent compatibility.

a-Lytic protease and subtilisin E are two enzymes that have been redesigned for nonaqueous biocatalysis. Both enzymes have been tested in mixtures of water and polar organic solvents for the effectiveness of the mutations. Substitution of surface charge residues by uncharged residues and the addition of residues with enhanced hydrogen bonding abilities had a positive effect on the enzymes' stability. Interestingly, the redesigned proteins, which performed so well in mixtures of water and organic solvents, have also been shown to function with improved efficiency in essentially anhydrous systems.

GOALS FOR THE FUTURE

The opportunities presented by nonaqueous biocatalysis are limited only by the imagination. In just one decade, this area of research has become one of the most active in science and engineering. While the technology has already been commercialized, full-scale adoption of nonaqueous enzymology will depend on the generation of a

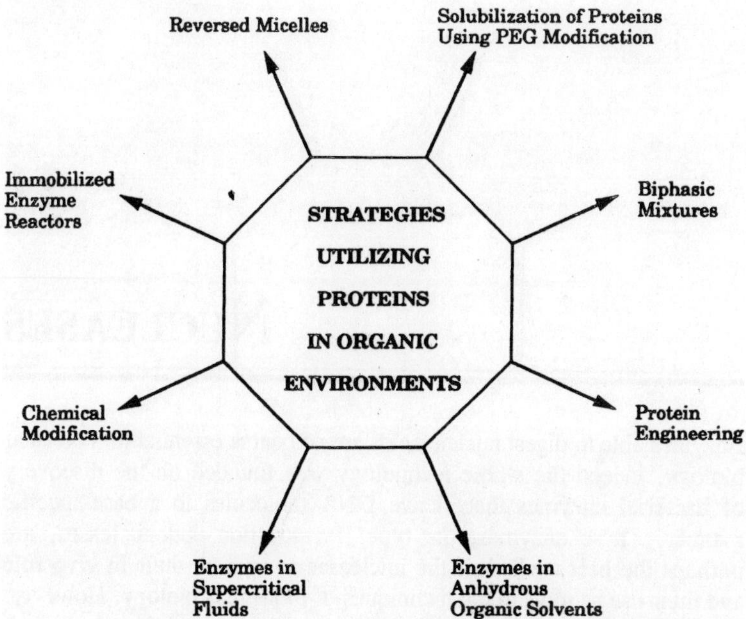

predictive model to correlate thc cffect of solvent properties to the binding and catalytic steps of an enzyme-catalyzed reaction. The design of bioreactors, and the development of upstream and downstream processing units for nonaqueous enzymology, are also central to further utilization of this core technology. The clear need for biocatalysts in industry should not obscure what the use of enzymes in extreme environments can teach us about the molecular basis of enzyme catalysis. Research is already demonstrating how the delicate interaction between environment and biological molecules serves to modulate function.

11

Nucleases

Enzymes able to digest nucleic acids are of course essential to molecular biology, indeed the whole technology was founded on the discovery of bacterial enzymes that cleave DNA molecules in a base-specific manner. These enzymes, the type II restriction endonucleases, are perhaps the best studied of the nucleases as to both their in vivo role and their use as tools in the techniques of molecular biology. However, the nucleases are ubiquitous in living organisms and function in all situations where partial or complete digestion of nucleic acid is required. These situations not only include degradation and senescence but also replication and recombination, although it must be noted that, to date, evidence for the involvement of nucleases in the latter two processes in eukaryotes is largely circumstantial. The significance of nucleases in the functioning of nucleic acids as the genetic material can be gaged however by considering that several enzymes implicated in DNA replication, recombination, and repair have integral exo- or endodeoxyribonuclease activity. For example, the 5'-3' and 3'-5' exonuclease activity of DNA polymerases and the endo-DNase activity of topoisomerases.

As well as the restriction endonucleases, various other nuclease enzymes have been used as tools in molecular biology, the purpose of this chapter is to give some background on the main deoxyribonucleases (DNases) and then to focus on the techniques in which they are used.

NOMENCLATURE

Anyone who has tried the isolation of a DNase enzyme will know that the presence of multiple types of nuclease activity makes this process fraught with difficulty. In this section, consideration will be given to the properties of the DNase enzymes with a view to understanding their nomenclature, which for the most part is somewhat confusing.

Table 11.1 : The Nomenclature of Nuclease

EC number	Reaction	Example
3.1.11	Exodeoxyribonucleases producing 5'-phosphomonoesters.	Exonuclease III EC 3.1.11.2
3.1 13	Exoribonucleases producing 5'-phosphomonoesters	Exo-RNase H EC 3.1.13.2
3.1.14	Exoribonucleases producing other than 5'-phosphomonoesters	Yeast RNase EC 3 1.14.1
3 1.15	Exonucleases active with either ribo- or deoxyribonucleic acids and producing 5'-phosphomonoesters	Venom exonuclease EC 3 1 15.1
3.1.16	Exonucleases active with either ribo- deoxyribonucleic acids and producing other than 5'-phosphomonoesters.	Spleen exonuclease EC 3 1.16 1
3.121	Endodeoxyribonucleases producing 5'-phosphomonoesters.	DNase I EC 3.1 21 1 Type II restriction DNases EC 3.121.4
3 1.22	Endodeoxyribonucleases producing other than 5'-phosphomonoesters.	DNase II
3 125	Site-specific endodeoxyribonucleases. specific for altered bases.	
3 126	Endoribonucleases producing 5'-phosphomonoesters.	RNase H EC 3 1.26.4
3.1 27	Endoribonucleases producing other than 5'-phosphomonoesters	RNase T 1 EC 31273
3.1 30	Endonucleases active with either ribo- or deoxyribonucleic acids and producing 5'-phosphomonoesters.	*Aspergillus* nuclease S 1 and Mung bean nuclease EC 3 1 30 1

3 1 31	Endonucleases active with either ribo- or deoxyribonucleic acids and producing other than 5'-phosphomonoesters.	Micrococcal nuclease EC 31311

Nucleases, although a large group in themselves, are part of a larger group of enzymes, the phosphodiesterases, which are able to catalyze the cleavage of phosphate-ester bonds. Schmidt and Laskowsi identified three types of nuclease enzymatic activity: DNases, ribonucleases (RNases), and exonucleases. On this definition, it is apparent that so-called DNases and RNases cleave their substrates endonucleolytically, i.e., at internal sites, and that this activity is distinct and separable from any exonuclease activity. In practical terms, this definition holds true in that an endo-DNase will not digest DNA molecules to completion, i.e., to nucleotide monomers; only when exonuclease activity is present will the digestion of DNA go to completion.

A second confusing element in the nomenclature of nucleases, and DNases in particular, is the presence of single-stranded DNases, e.g., mung bean nuclease and nuclease S1 from *Aspergillus*. These enzymes, although having high specificity for single-stranded DNA molecules, will, at high concentrations and in preparations not purified to homogeneity, digest native (double-stranded) DNA molecules albeit at reduced rates. For an example of this, Weir and Bryant (3) have isolated a nuclearlocated DNase from the embryo axes of pea that has a low, but measurable activity on native DNA but rapidly catalyzes the hydrolysis of heat-denatured DNA. It is not known so far whether these two activities are separable, but evidence from similar enzymes suggests that these activities are part of the same protein molecule. DNases then, tend to be classified as to "what they do best"; e.g., the DNase of Weir and Bryant would be called a single-strand specific endo-DNase. In the following discussion, the examples are from the DNase class of nucleases, however all the points considered can be equally applied to the RNases.

Criteria Used for Classification

Exo-vs Endonucleolytic Cleavage

Exo-DNases cleave from the ends of DNA molecules releasing phosphomononucleotides. Cleavage can be either in the 3' to 5' direction releasing 5' phosphomononucleotides or in the 5' to 3' direction to yield 3' phosphomononucleotides. An example of a widely used exonuclease is exonuclease III from *Escherichia coli* (EC

3.1.11.2), which will digest one strand of a double-stranded DNA molecule from a 3' overhang or blunt end. This property has been used to produce bidirectional or unidirectional nested deletion of templates for sequencing.

Endo-DNases cleave at internal phosphate bonds. Cleavage of double-stranded DNA substrates can be by a "single-hit" or a "doublehit" mechanism or by a combination of both. Essentially this means that the enzymes can either cleave the two strands of the DNA molecule at points opposite or at sites on the two strands that are well away from each other. The scission of the molecule will take place at a relatively faster rate in the former case as compared to the latter. The prime example of an endo-DNase is pancreatic DNase (DNase I, EC 3.1.21.1). Under optimal conditions this enzyme uses a double-hit mechanism for cleavage of substrates.

Base Specificity at or Near the Site of Cleavage

None of the eukaryotic enzymes so far isolated appear to have such specificity, however there is evidence that enzymes with optimal activity on single-stranded DNAs will preferentially cleave at A-T rich sites in native DNA molecules. As already mentioned, the Type II restriction endonucleases have absolute specificity for a group of bases at or near the cleavage site.

Site of Cleavage

The site of cleavage can be on either side of the phosphate bond leading to a 5' or a 3' monoesterified product. No enzymes have been isolated that can split the internucleotide bond on either side. This property is particularly important if the DNA molecule is to be subsequently made blunt-ended for a ligation experiment. DNA molecules left with a 5' overhang from a staggered cut are usually filled in with the Klenow fragment of *E. coli* DNA polymerase, whereas those with a 3' overhang have the overhang cleaved back with the exonuclease activity of T4 DNA polymerase to create blunt ends.

Cleavage of Native or Single-Stranded DNA

Nucleases tend to have a "preference" for cleavage of single-stranded DNA or double-stranded DNA substrates. In general single-strand specific DNases, such as nuclease SI *from Aspergillus* (EC 3.1.30. 1), will digest native DNA if the enzyme concentration is high. Nuclease S1 is used to analyze the structure of DNA-RNA hybrids and in cDNA synthesis where it opens the hairpin loop generated during the synthesis.

General

DNases have also in the past been classified according to their pH optima and requirements for the presence of metal ions; there are two major mammalian DNases-one working at neutral pH the other in acidic conditions-and both require Mg ions. Other nuclease enzymes exist that require Ca^{2+} ions (pea nuclear DNase) or Zn^{2+} ions (nuclease S1). In addition to the aforementioned properties, it is important to note that many crude preparations of DNase exhibit nonspecificity for the sugar moiety of nucleic acids, i.e., they will cleave both RNA and DNA. In these cases it is obviously essential to remove contaminating RNase activity before the enzyme is used to remove DNA from a preparation of RNA.

From the foregoing discussion, it can be seen that the nucleases are a complex group of enzymes. However, when it comes to their use as tools in molecular biology, the situation is very much simplified as only a handful of enzymes are used routinely in experimental protocols. The enzymes DNases I and II, exonucleases, nuclease Sl, *Bal31,* and RNase will be discussed in much more detail in the following sections with special emphasis on the experimental protocols for their use in molecular biology techniques.

DEOXYRIBONUCLEASE

The term DNase refers to an enzyme that endonucleolytically cleaves DNA molecules. This chapter deals with those DNases that preferentially catalyze the hydrolysis of double-stranded DNA (ds DNA) and that have found a use in the various techniques employed in molecular biology. The Type II restriction endonucleases obviously fall into this category, however they are a large subject on their own and therefore are dealt with in a separate section of this book. DNases acting preferentially on singlestranded DNA (ss DNA) substrates are also dealt with in another section of this chapter.

Despite the ubiquitous nature of nucleases in living organisms, it is perhaps surprising to find that, apart from the Type II restriction endonucleases, only *one ds* DNA DNase enzyme is used routinely in molecular biology. The reasons for this are twofold. It is very difficult to isolate a DNase in a form that is completely free of accompanying RNase and exonuclease contamination and indeed these activities may be part of the same enzyme molecule. Obviously, it would not be desirable to remove DNA from a precious RNA preparation with a DNase that has an integral RNase activity. The second reason is that the first DNase isolated from tissues, DNase I (or pancreatic DNase), performs all the tasks very well and can be isolated readily in a pure form; why use or look for another enzyme? DNase I has been commercially available for at least 25 years, most of the work on the properties of the enzyme being done in the 1960s and, to date, no DNase has been found to replace it in the dual tasks of complete degradation of "nuisance material" DNA and the partial hydrolysis of DNA molecule in such techniques as nick translation.

In the following sections, the properties of DNase I are discussed in detail with special emphasis directed toward comparing the activity of DNase I with that of other types of DNase, most notably DNase II.

DNASE I

Reaction

Bovine pancreatic DNase, or more usually, DNase I, catalyzes the hydrolysis of ds DNA molecules but, at high concentrations of enzyme, ss DNA will also be digested. "Complete" hydrolysis results in the formation of small oligonucleotide products that are resistant to further cleavage, but are acid soluble *(see* section on assay of DNase activity); cleavage results in the formation of 5' monoesterified products. Cleavage of DNA substrate with DNase II (spleen DNase or acid DNase), in contrast, results in the formation of 3' monoesterified products.

pH Optimum

Possibly the major reason that DNase I is preferred to DNase II is that DNase I has optimum activity in the region of pH 7-8, whereas DNase II, as its alternative name describes, has a pH optimum in acidic condition: pH 4.2-5.5.

Activators and Inhibitors

DNase I has an absolute requirement for divalent metal cations. The most commonly used is Mg^{2+}, however Mn^{2+}, Call, Co^{2+}, and Zn^{2+} will also activate DNase I. Concentrations of Mg^{2+} above approx 50 mM become inhibitory, which is not the case for Co^{2+} and Mn^{2+}. Monovalent metal ions are also inhibitors of the enzyme activity. In the presence of Call, Mg^{2+} has a synergistic effect, i.e., the rate of hydrolysis of DNA in the presence of both ions is more than the sum of the rates of hydrolysis of DNA in the presence of each ion separately. A total of 0.1 mM Ca^{2+} is sufficient to give this enhanced reactivity in the presence of 10 mM Mg^{2+}, however the rate of hydrolysis of DNA is still greatest in the presence of Mn^{2+} ions.

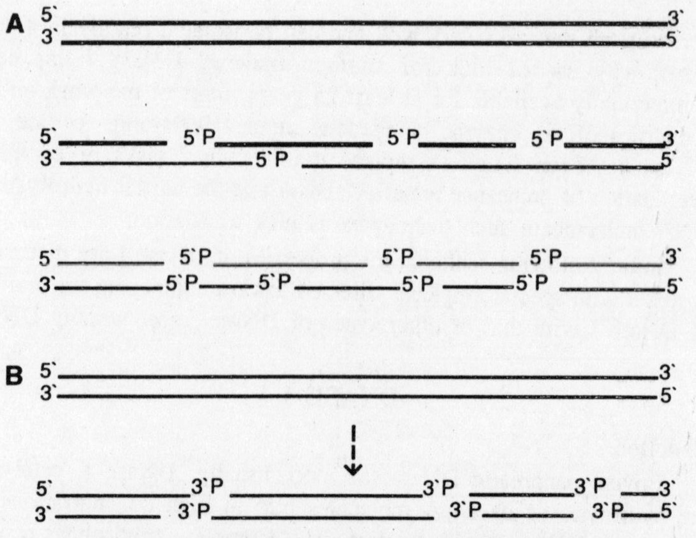

Figure 11.1 : A. Double-hit mechamsm of DNase I m presence of Mg^{2+} ions B. Smglehit mechanism of DNase II.

Apart from the monovalent cations, there is no general inhibitor of DNase I such as those available for the inhibition of RNases. The only real way to combat DNase activity, which may be a worry during DNA extractions, is to do the extraction as quickly as possible and at low temperatures. Inclusion of EDTA in the extraction buffer is a good idea, but note that Jones and Boffey recently discovered a DNase in the leaves of wheat seedlings that appears to be stimulated in the presence of EDTA.

KINETICS

Using a variety of methods, including light-scattering, viscometry, and sedimentation analysis, it can be shown that there are two different types of mechanism for the cleavage of ds DNA substrates by DNase. DNase I, at low concentrations and under the usual assay conditions, inserts nicks at random points in each strand of the DNA at points away from each other. This is termed a "double-hit" mechanism; complete scission of the molecule will not occur until two nicks are opposite. Monitoring of the reaction therefore will not indicate the presence of DNA molecules of intermediate size until after a lag phase. In contrast, DNase II was shown to cleave high-mol-wt DNA substrates on both strands at points opposite to each other resulting in the complete scission of the molecule. This process was termed a single-hit mechanism, as scission of the DNA occurred from a single encounter with the enzyme molecule; during a digestion, intermediate size molecules will appear immediately. An interesting feature of DNase I is that the cleavage mechanism can be altered from double-hit to a DNase II-like single-hit mechanism by using high concentrations of the enzyme or by altering the divalent cation from Mg^{2+} to Mn^{2+} or Co^{2+}. Using ss DNA as substrate, Melgar and Goldthwaite (2b) showed that in the presence of $Mn^{2}+$ ions, DNase I had vastly increased V_{max} as compared to the activity in the presence of Mg^{2+}, whereas there was little change in the K_m. The increased rate of hydrolysis in the presence of Mn^{2+} could of itself lead to the formation of intermediate size fragments in the short periods observed without there being a change in the actual mechanism of cleavage of a double-stranded substrate; the greater the number of random nicks, the greater the likelihood of there being scission of the molecules. If viscosity measurements are used to follow the progress of a DNase digestion, a relationship can be obtained when the log of a function of the change in viscosity of the DNA solution is plotted against log time. The slope of the resulting line, *n*, gives an indication of the mechanism of the reaction: A value of approx 1.0 indicates single-hit kinetics, whereas a value between 1 and 2 indicates a predominantly double-hit mechanism. When the hydrolysis of ds DNA by DNase I in the presence of Mn^{2+} is monitored by viscometry at low temperature, i.e., at low rate of hydrolysis, a value of $n = 1.16$ is obtained, indicating that the reaction is predominantly of the single-hit kind. Monovalent cations also lower the rate of hydrolysis of ds DNA by DNase I in the presence of Mn^{2+}, and under these conditions the *n* value changes from approx 1.0 (single-hit mechanism) to values approaching 2.0 (double-hit mechanism). The inhibition of DNase I by monovalent cations is probably a result of competition for effector sites that directly or indirectly affect the active site of the enzyme

The process by which Mn^{2+} promotes the hydrolysis of ds DNA by DNase I to switch from a double-hit mechanism to a single-hit mechanism is not known but may involve either (1) the ability of Mn^{2+} to promote the binding of two DNase I molecules at opposite sites on the two DNA strands (it is known that active DNase II is dimeric); or (2) the ability of Mn^{2+} to enable the DNase I enzyme molecule to flip from the hydrolyzed strand to the opposite strand, which is then also cleaved.

SPECIFICITY

There is evidence that certain deoxynucleotide sequences can be hydrolyzed in the presence of Mn^{2+} but not in the presence of Mg^{2+}.

For example, poly (dG:dC) can be hydrolyzed by DNase I with Mn 2+ but not with Mg^{2+}. This resistance is not attributable to secondary structure of the double-helix, as DNase I in the presence of Mg^{2+} is not able to hydrolyze the dC strand of a polymer consisting of dl:dC. Therefore, in addition to altering the kinetics of the reaction, different metal ions can also place qualitative constraints on hydrolysis of certain residues.

EXPERIMENTAL PROCEDURES

DNase I is the enzyme of choice for all molecular biology techniques (apart from restriction digestions) that require a double-stranded DNase enzyme. It is readily available in pure form, i.e., exonuclease and RNase free, from a number of commercial sources, has high activity on ds DNA, and at high concentrations, will also cleave ss DNA. The reaction does not go to completion, to produce deoxynucleotide monomers, but a certain limit size is reached, after which no further digestion takes place. DNase I cleavage leads to the formation of 5' monoesterified products and the enzyme has optimum activity in the pH 7.0 region. DNase I is inhibited by monovalent cation but has an absolute requirement for divalent metal ions. The nature of the cation affects the mode of action of the enzyme and its ability to cleave at certain deoxynucleotide residues. The enzyme has optimal activity in the presence of Mn^{2+} cations, but it is noteworthy that molecular biologists use Mg^{2+} as the cofactor. This is probably attributable to the fact that many of the enzymes used in conjunction with DNase also require Mg^{2+} as cofactor, e.g., DNA polymerase I in the nick-translation protocol, and that many of the techniques require only limited digestion of the DNA.

Assay of DNase Activity

Many methods have been used in the past to assay DNase activity: Melgar and Goldthwaite used a somewhat complicated method whereby T4 DNA was labeled with ^{32}P and embedded in a polyacrylamide gel. After exposure of the gel to DNase, the gel was pelleted and the amount of radioactive DNA fragments released into the supernatant was measured. More recent methods have again used radiolabeled substrates, but this time the rate of hydrolysis is monitored by precipitating the DNA fragments produced out of aqueous solution with trichloroacetic acid (TCA) in the presence of a carrier and then measuring acid-soluble or acid-insoluble radioactivity. For most molecular biologists however, all that is required is an assay method to show whether DNase is responsible for the loss of DNA from a preparation, and if so, to show where the contamination is coming from. This is best accomplished by using a covalently closed circular (ccc) plasmid as substrate in a component analysis. In the following example, the components of a restriction enzyme digestion of a plasmid are to be analyzed.

1. Set up several reactions depending on the number of components to be tested, remember that the controls are all important.
   a. ccc plasmid in sterile distilled water (500 ng in 10 μL)
   b. ccc plasmid in buffer
   c. ccc plasmid in buffer + bovine serum albumin (BSA)
   d. ccc plasmid in buffer + enzyme
   e. ccc plasmid in buffer + BSA + enzyme
2. Incubate at 37°C for 30-60 min.
3. Run digests on an agarose gel.
4. Stain with ethidium bromide and view under UV illumination.

Endo-DNase activity is indicated by excessive conversion, as compared to the controls, of ccc plasmid to open circles or to linears. Usually endo-DNase contamination will be accompanied by exonuclease activity, so there will be complete degradation of the plasmid to a smear of material down the gel.

Removal of DNase Activity

DNase enzymes are heat labile and can be removed from solutions by autoclaving and from glassware and the like by baking or autoclaving. To make sure that solutions of RNase I are DNase free, boil the RNase solution for 15 min (RNase I is a remarkably stable enzyme) and cool to room temperature. Dispense into aliquots and store frozen.

As already mentioned, during the preparation of DNA, endogenous DNases can be inhibited by the inclusion of EDTA in the homogenization buffer. It was also pointed out that a DNase activity had recently been discovered that was apparently stimulated by EDTA. The prevalence of this type of nuclease activity is not known, but if it is suspected, damage can be limited by performing extractions at low temperatures and proceeding to the protein removal step as rapidly as possible

Removal of RNase from DNase Solutions

RNase-free preparations of DNase I are now commercially available, however some of these still may not be entirely satisfactory.

Various column chromatography methods have been used to remove RNase activity from DNase solutions and an effective method for disabling the RNase A enzyme has been described by Zimmerman and Sadeen and modified by Gurney and Gurney (8). This method depends on the alkylation of a histidine in the active site of RNase A (a likely contaminant of DNase I as they are derived from the same tissue) using sodium iodoacetate. A brief protocol is outlined in the following:

1. Use autoclaved micropipet tips and glassware.
2. Dissolve 10 mg commercial DNase in 2 mL of 2.5 mM HCI.
3. Dialyze the solution for a few hours at 40°C against I L of 2.5 *MM* HCl with stirring, then overnight against I L of fresh 2.5 mM HCl again at 4°C with stirring.
4. Freshly prepare *1M* sodium iodoacetate. Mix 2.5 mL of 0.2M sodium acetate, pH 5.3, 2.0 mL of dialyzed enzyme, and 7.5 ml, of *1M* sodium iodo acetate. Incubate the mixture at 55°C for 60 mm A precipitate will form.
5. Dialyze the solution overnight against 1 L of 2.5 mM HCl at 4°C.
6. Centrifuge solution at 10,000g for 30 min at 0°C.
7. Carefully pipet off the supernatant into a sterilized tube or bottle with a tight fitting cap.
8. The protein concentration should be 3-4 mg/mL. The DNase should be stored at 4°C, the activity should be stable (unfrozen) for at least 2 yr.

RNases can also be inhibited by a number of compounds including vanadyl ribonuclease complex, heparin, and their own specific ribonuclease inhibitors. RNases are dealt with in much more detail elsewhere in this volume.

"DEGRADATIVE" USES OF DNASE I

Stock solutions of DNase I should be made up at 2 mg/mL in 10 MM Tris-HCI, pH 7.5, 50 mM NaCl, 10 mM $MgC1_2$, 1 mM dithiothreitol, and 50% glycerol, and stored frozen in small aliquots. Removal of DNA from preparations can be accomplished by incubation with 2050 pg/mL DNase I in the presence of 50 mM Tris-HCI, pH 7.5, 10 MM $MgCl_2$, at 37°C for 60 min.

A method for preparing random oligonucleotide fragments from calf thymus DNA for use in the oligolabeling protocol.

Nick Translation

The alternative method for producing radiolabeled probes other than oligolabeling is nick translation. In this method DNase I is used to nick the DNA to be labeled creating free 3'OH groups. *Escherichia coli* DNA polymerase I catalyzes the addition of nucleotides to the 3'OH termini and the 5'-3' exonuclease activity of the same enzyme removes nucleotides from the 5' end thus translating the nick. Linear, supercoiled, nicked, or gapped circular ds DNA molecules can be labeled to specific activities of > 10^8 cpm/pg with ^{32}P labeled deoxynucleotide triphosphates. Meinkoth and Wahl pointed out that, at saturating levels of nucleotide triphosphates, the size of the labeled fragments is determined by the DNase concentration: Fragments of 500-1500 nucleotides long produce optimal signal:noise ratios in the subsequent hybridization. The necessary concentration of DNase required can be determined by doing a titration using from 0-100 pg/mL DNase in the reactions and running the labeled DNA on acrylamide gels. DNase concentration, which gives approx 30% incorporation of label, is optimal.

DNA FOOTPRINTING BY DNASE I PROTECTION

DNase I has been used in the past to dissect the structure of chromatin and to isolate nucleosomes, but in recent years molecular biologists have become interested in the proteins that bind to DNA and possibly regulate the expression of genes. The isolation and characterization of such transactivating factors can be accomplished by using the powerful technique of DNA footprinting. In this technique the DNA of interest is labeled at either the 3' or 5' end and protein is bound to the DNA; the nucleoprotein complex is then subject to limited digestion with DNase I. Under the conditions used, DNase will exhibit double-hit kinetics and, therefore, will insert random single-strand nicks, however, cleavage will be inhibited if protein is already bound to the DNA. The DNA is then purified, denatured, and run on a sequencing polyacrylamide gel. Autoradiography of the labeled DNA in the gel should show a ladder of fragments representing cleavage at approximately every base for the control with no protein present. However, if protein has been bound to a particular sequence, there will be no cleavage by DNase I and there will be a gap in the ladder. This gap is known as a footprint; the sequence of the footprint can be determined if a sequencing reaction is run alongside the DNase I protection assay. Also note that Stratagene now markets a DNA footprinting kit using the DNase I protection method. Exonuclease III protection has also been used for DNA footprinting, but in this technique the DNA must be labeled at the 5' end.

DNA POLYMERASES

DNA polymerases catalyze the synthesis of DNA molecules from monomeric deoxynucleotide triphosphate units. This definition encompasses those enzymes classed as DNA-dependent DNA polymerases (which require both a DNA template strand and a DNA primer to wh ich the monomeric units can be added), reverse transcriptases (RNA-dependent DNA polymerases that utilize an RNA template), and terminal deoxynucleotidyl transferases (which require no template).

DNA-dependent DNA polymerases are a class of enzyme that is essential for the replication and maintenance of all organisms. All perform essentially the same reaction, the addition of monomeric units to synthesize a complementary copy of an existing DNA template. This activity can be used in molecular biology for a wide range of techniques. These include the labeling of DNA molecules, terminally or throughout the sequence, for use as probes or for sequence determination, the modification of DNA termini for facilitating subsequent manipulations, the production of double-stranded cDNA, the extension of mismatched sequences for site-directed mutagenesis, the extension of synthetic DNA sequences for gene synthesis, and the specific amplification of limiting quantities of DNA for analysis.

The choice of enzyme depends on the reaction conditions employed and the desirability of other inherent enzymatic activities. The range of enzymes available is rapidly expanding, ensuring a concomitant increase in the applications of DNA polymerases in molecular biology.

Prokaryotic DNA Polymerases

The classical *E. coli* DNA polymerase family consists of DNA polymerases I, II, and III (Pol I, Pol II, and Pol III). Pol III is largely responsible for chromosomal replication, and it initiates the growth of both DNA chains on RNA primers. Pol I "fills in" the gaps left by the intermittent nature of Pol III activity on the 3'-5' strand. Apart from this activity, the major role of Pol I is in DNA repair rather than replication. The role of Pol II is still largely unclear.

The three enzymes have widely differing properties. Pol I, which is the most widely used in molecular biology, is described in the next section. Pol II has a polymerase activity of only 5% that of Pol

I and has no 5'-3' exonuclease activity. It can only act on duplexes with gaps of < 100 bases and cannot replicate long single-stranded stretches from short primers. Pol III has a similar template requirement to Pol II, but is 300 times more active. Pol III holoenzyme consists of at least 13 components, although various smaller complexes are active in vitro.

In addition to the *E. coli* enzymes, DNA polymerases from other bacteria are now attaining more prominence in molecular biology, particularly those from thermophilic organisms. Other useful DNA polymerases have been isolated from bacteriophage-infected *E. coli* cells, especially the enzymes encoded by bacteriophages T4 and T7.

Eukaryotic DNA Polymerases

Eukaryotic organisms possess three main DNApolymerases, Pol α, Pol β, and Pol γ. Pol α is considered to be the major DNA polymerase and is responsible for chromosomal replication. The function of Pol β is unclear, whereas Pol γ, being located in both the nucleus and the mitochondria, is believed to be involved in mitochondrial DNA replication. The three enzymes all lack 3'-5' exonuclease activity, which has since been detected on a fourth DNA polymerase, Pol δ. This finding supports the idea that eukaryotic DNA polymerases work in concert during DNA replication.

DNA POLYMERASE I

DNA polymerase I, Pol I, or Kornberg polymerase is the product of the *pol A* gene of *E. coli*, which has now been cloned to yield an overproducing strain.

Enzyme Data

The enzyme comprises a single polypeptide chain of mol mass 109,000 Da, with various enzymic activities. These are a 5'-3' polymerase, a 3'-5' exonuclease (proofreading activity), a 5'-3' exonuclease, and a phosphate-exchange activity. The polymerase activity catalyzes the formation of a complementary DNA strand to an existing single-stranded template, by the extension of a DNA primer possessing a 3'-hydroxyl terminus. The latter may be provided merely by a short gap or even a nick in the DNA duplex. The presence of these multiple activities allows the use of DNA polymerase I to perform several complex manipulations.

DNA polymerase I requires the presence of magnesium ions, as do all DNA polymerases. (In some cases, manganese ions may substitute, but reaction rates are lower, and there is a loss of fidelity.)

The pH optimum for the enzyme is 7.4, and sulfhydryl reagents are required.

Enzymic activity is measured by the polymerization of labeled nucleotides into double-stranded DNA using poly-dAdT as both template and primer. One unit of activity catalyzes the incorporation of 10 nmol into an acid precipitable form in 30 min at 37°C. DNA Polymerase I can be obtained at a purity of 5000 U/mg.

Uses of DNA Polymerase I

There are several uses of DNA polymerase I in molecular biology. First, it is used for the incorporation of labeled nucleotides into DNA probes by nick translation. This requires the action of deoxyribonuclease *I* to introduce nicks into the duplex with exposed 3'-hydroxyl termini to act as a substrate for the DNA polymerase I. 5'-3' Polymerization occurs concomitantly with 5'-3' exonuclease activity, which has the effect of physically translating the "nick" along the DNA molecule and allowing the introduction of labeled nucleotides into the newly replaced DNA strand. Reaction conditions are such that one complete exchange of the DNA duplex is achieved in this manner. DNA polymerase I is the only common DNA polymerase that can carry out this coupled reaction by virtue of its 5'-3' exonuclease function.

Disadvantages of nick translation with regard to other methods of probe labeling include the facts that the specific activity that may be achieved is not high, and that at temperatures above 20°C artifactual sequences are produced, such as "snapback DNA," where the newly synthesized strand disassociates from the template and is then itself copied to form inverted DNA sequences.

Another use of DNA polymerase I is in repair synthesis, where the enzyme is used to "fill in" protruding 5' termini. End-labeling can also be achieved by removal of 3' protruding termini by the 3'-5' exonuclease activity and then subsequently filling in the 3' recess so formed.

DNA polymerase I can also be used in cDNA synthesis, where it is employed in the synthesis of the second DNA strand. The enzyme has an inherent RNase H activity that is of value here. However, cDNA synthesis is largely performed by DNA polymerase I large (Klenow) fragment or reverse transcriptase.

EXPERIMENTAL PROCEDURES

DNA polymerase I is best stored in 50 mM potassium phosphate, pH 7.0, containing 250 EM DTT and *50%* glycerol. At –20°C, it is stable for over 6 mo.

Protocol for Nick Translation with DNA Polymerase I

Materials Required

1. Probe DNA.
2. Nick translation buffer:
   - 500 *mM* Tris-HCI, pH 7.8
   - 50 *mM* $MgCl_2$
   - 100 *mM* β-Mercaptoethanol
   - 100 μg/mL Bovine serum albumin
3. 200 MM dGTP.
4. 200 *μM* dCTP.
5. 200 μM dTTP.
6. 0.1 μg/mL DNase I (prepare by serial dilution of 1 mg/mL stock).
7. DNA polymerase I (diluted to 2 U/μL in storage buffer).
8. 0.5M EDTA, pH 8.0.
9. THE buffer:
   - 10 *mM* Tris-HCI, pH 8.0
   - 100 *mM* NaCl
   - 1 *mM* EDTA
10. TE buffer:
    - 10 *mM* Tris-HCI, pH 8.0
    - 1 *mM* EDTA
11. Sephadex G-50 equilibrated in THE buffer.
12. α-^{32}P dATP (10 mCi/mL, 3000 Ci/mmol).
13. Sterile distilled water.

Equipment Required

1. Water bath (15°C).
2. Water bath (70°C) or dry heating block.
3. Bench-top centrifuge.
4 Sterile glass wool.
5. Sterile 1.5-mL microfuge tubes.
6. Pipets and sterile pipet tips.
7. Ice bucket.

Protocol

1. In a sterile tube on ice mix the following:
   - Probe DNA (up to I μg)
   - 5 μL Nick translation buffer
   - 200 μM dGTP, dCTP, dTTP (5 μL each)
   - 5 μL α-^{32}P dATP (50 μCi, 20 μM)
   - 2 μL DNase I (200 pg)
   - 1 μL DNA polymerase I (2 U)
   - Sterile distilled water to 50 μL
2. Incubate at 15°C for 60 min.
3. Stop the reaction by addition of 2 μL of 0.5M EDTA, pH 8.0, or by heating to 70°C for 10 min.
4. Purify the labeled probe from unincorporated nucleotides by a spun column method. Prepare a column of Sephadex G-50 in a 200-μL pipet tip plugged with glass wool. Wash the column with 500 μL TNE buffer, by supporting it in a microfuge tube and spinning at 2000g for a few seconds. Load the nick translation reaction onto the column, and spin for 30 s. Elute the labeled probe with a futher 100 μL of TNE (or TE) buffer and a further spin. Pool the run-through and eluate, and then discard the column containing the unincorporated nucleotides.

KLENOW FRAGMENT

Enzyme Data

By treatment of DNA polymerase I with the protease subtilisin, a polypeptide of mol mass 35,000 Da is removed that carries the 5'-3' exonuclease activity. The remaining moiety, of mol mass 76,000 Da, retains all the other enzymic activities and is commonly termed the DNA polymerase I large fragment or Klenow fragment.

Uses of Klenow Fragment

It is impossible to use Klenow fragment for nick translation because of the lack of a 5'-3' exonuclease activity. However, this limitation is turned to advantage in other applications of the enzyme by rendering the polymerization activity more controllable.

The primer extension activity of Klenow fragment has made it widely used in dideoxynucleotide sequencing methods, whereby DNA templates up to 350 bases long can be sequenced at a rate of 45 nucleotides/s. Several different DNA polymerases are commonly used for sequencing reactions. Each has distinct abilities when encountering particularly "difficult" regions of the template sequence.

The enzyme can also be used to produce labeled double-stranded DNA probes (actually only labeled in one or the other of the two strands) by the extension of oligomeric primers of random sequence. The oligolabeling technique produces probes of a much higher specific activity than nick translation, with over 55% of the labeled nucleotide incorporated in 30 min at 37°C. Although the entire probe sequence is represented in the labeling mix, individual molecules are commonly only a few hundred bases in length, which may be a disadvantage for some applications.

Single-stranded (strand-specific) probes can also be prepared by end-labeling of protruding, flush, or recessed 3' termini. This is achieved by removal of the 3' nucleotides by means of the 3'-5' exonuclease activity (which is more active on single-stranded than double-stranded DNA), to form a recessed 3' end. Following the addition of high concentrations of deoxynucleotide triphosphates (one or more of which may be labeled), the polymerase activity resynthesizes the excised region. This process has advantages over nick translation in that the labeled probe produced is strand-specific, and there is no production of hairpin-like artifacts.

This reaction can of course be performed with unlabeled deoxynucleotides to produce flush or modified termini for subsequent cloning reactions. The fill-in reaction can be controlled by the withholding of one particular deoxynucleotide to generate novel cohesive ends. Excess Klenow fragment will also yield only partially filled termini because of the 3'—5' exonuclease activity.

Klenow fragment is also used for elongating oligomeric primers in site-directed mutagenesis to produce double-stranded DNA. However, the enzyme tends to displace mismatched oligomers and so remove the intended mutation. T4 DNA polymerase is perhaps more useful in this context.

The enzyme can also be used for the synthesis of the second DNA strand during cDNA synthesis. Finally, Klenow fragment was originally used in the polymerase chain reaction, before the introduction of thermostable DNA polymerases permitted the process to be automated.

EXPERIMENTAL PROCEDURES

Klenow fragment has similar physical properties to Pol I and should be stored under similar conditions. The following sections describe experimental protocols for two techniques utilizing Klenow fragment: radiolabeling DNA by extension of random primers and in-filling of cohesive termini.

Protocol for Labeling with Random Hexameric Primers

Materials Required

1. Probe DNA.
2. Reaction buffer:
   - 450 *mM* HEPES, pH 6.6
   - 50 *mM* $MgC1_2$
   - 10 *mM* Dithiothreitol
   - 2 *mg/mL* Bovine serum albumin
   - 2.5 *mM* dGTP
   - 2.5 *mM* dCTP
   - 2.5 *mM* dTTP
3. Klenow fragment (diluted to 2 U/μL in storage buffer).
4. α-^{32}P dATP (10 mCi/mL, 3000 Ci/mmol).
5. Sterile distilled water.

Equipment Required

1. 1.5-mL Microfuge tubes.
2. Microfuge.
3. Boiling water bath or dry heating block.
4. Water bath (37°C).
5. Pipets and sterile pipet tips.
6. Ice bucket.

Protocol

1. Dilute 50 ng probe DNA to 34 μL with sterile water.
2. Denature by boiling for 3 min. Spin down condensation and stand on ice.
3. Add 10 μL of reaction buffer.
4. Add 5 μL (50 μCi) α-^{32}P dATP.
5. Add 1 μL (2 U) Klenow fragment.
6. Incubate at 37°C for 30 min or at room temperature for 1-2 h.
7. Add 200 μL sterile water, and terminate the reaction by boiling for 3 min.
8. Stand the reaction on ice and use (unpurified) within 10 min.

Protocol for In-Filling Cohesive

Materials Required

1. Target DNA.
2. Reaction buffer:
   - 70 *mM* Tris-HC1, pH 7.5
   - 70 *mM* $MgC1_2$
   - 10 *mM* Dithiothreitol
   - 250 *μM* dATP
   - 250 *μM* dGTP
   - 250 *μM* dCTP
   - 250 *μM* dTTP

   N.B. Alternatively, the in-filling restriction endonuclease buffers.
3. 500 *mM* NaCl.
4. Klenow fragment (diluted to 0.5 U/μL
5. Sterile distilled water.

Equipment Required

1. 1.5-mL Microfuge tubes.
2. Water bath (25°C)
3. Water bath or dry heating block (70°C).
4. Pipets and sterile pipet tips

Protocol

1. Dilute I μg of target DNA to 44 μL with distilled water.
2. Add 5 μL of reaction buffer.
3. Optionally, add NaCl to a concentration of 0-50 mM, reducing the volume of water use in step I *pro rata.*
4. Add I μL (0.5 U) of Klenow fragment.
5. Incubate at 25°C for 15 min.
6. Terminate the reaction by incubating at 70°C for 10 min.

T4 DNA POLYMERASE

T4 DNA polymerase is produced by the E. *coli* bacterium when infected with bacteriophage T4.

Enzyme Data

It is a monomeric enzyme of mol mass 114,000 Da, that will sequentially add mononucleotide units to a DNA primer by means of its 5'-3' polymerase activity. The enzyme cannot utilize nicked DNA as a substrate, since its lack of a 5'-3' exonuclease activity prevents it from displacing the existing downstream DNA strand. The enzyme has a highly active 3'-5' proofreading exonuclease activity, 200-fold more active than that of Klenow fragment, which again is more active against single- than double-stranded DNA.

Uses of T4 DNA Polymerase

T4 DNA polymerase is used to label 3' termini in the same manner as Klenow fragment. The reaction can be performed at 37°C or at room temperature, but incubation at 12°C maximizes the polymerase activity over that of the exonuclease.

The exonucleolysis/in-filling process can of course be performed in the absence of labeled deoxynucleotide triphosphate to convert cohesive or "ragged" termini to "polished" flush ends suitable for cloning. A further use of T4 DNA polymerase is in the process of "gap-filling" during site-directed mutagenesis. This entails the conversion of a partially single-stranded molecule to a fully double-stranded entity. For this application, it is usual to include the T4 gene 32 protein, which stimulates DNA synthesis.

EXPERIMENTAL PROCEDURES

T4 DNA polymerase will function in most restriction endonuclease buffers. An alternative reaction buffer is 50 *mM* Tris-HCI, pH 8.5,15 mM ammonium sulfate, 7 mMMgC1_2, 0.1 *mM* EDTA, 10 *mM* β-mercaptoethanol, 33 μM dNTPs, and 200 p.g/mL BSA. The enzyme requires Mg^{2+} (at least 6 mM) for activity. Its pH optimum is in the range 8.0-9.0, exhibiting 50% activity at pH 7.5 and 9.7. The reaction temperature should be 37°C.

The enzyme is best stored in 10 mM Tris-HCI, pH 8.0, 100 mM NaCl, 2.5 *mM* MgC1_2, 0.5 mM EDTA, 2 *mM* DTT, and 50% glycerol. It is stable if stored undiluted in this buffer at -20°C for several months.

T7 DNA POLYMERASE

This DNA polymerase is also produced by bacteriophage-infected *E. coli,* although it is now obtained from a cloned gene.

Enzyme Data

It is the only common DNA polymerase that consists of more than one subunit. The functional enzyme is composed of the T7 gene 5 protein (mol mass 80,000-85,000 Da), which carries all the enzymic activities, plus the *E. coli* thioredoxin protein (mol mass 12,000 Da) required to associate with the template.

The enzyme has a very active 5'-3' polymerase activity that can incorporate 300 nucleotides/s. It is also the most highly processive of all the DNA polymerases in that it can extend a primer for several kilobases without the enzyme dissociating from the template. The enzyme is also very tolerant of nucleotide analogs, which renders it very useful for DNA sequencing and oligolabeling techniques.

T7 DNA polymerase lacks a 5'-3' exonuclease activity, but like T4 DNA polymerase, it has a potent 3'-5' exonuclease. In this case, the activity is 1000-fold that exhibited by Klenow fragment. The 3'-5' exonuclease activity can be inactivated by incubation with a reducing agent, oxygen, and low concentrations of Feel. This modification does not affect the polymerization ability of the enzyme. Chemical modification removes 99% of the exonuclease activity, and total removal has been achieved by genetic engineering. These modified versions of T7 DNA polymerase are readily available commercially.

Uses of T7 DNA Polymerase

A major use of native or modified T7 DNA polymerase is in DNA sequencing when fragments of over 1,000 bases can be analyzed from one reaction. The enzyme is also useful for other applications when extension of a primer over long stretches of template are required. The other uses of this enzyme include the rapid end-labeling of DNA probes and the extension of primers during site-directed mutagenesis.

EXPERIMENTAL PROCEDURES

A suitable reaction buffer for T7 DNA polymerase is 40 mM TrisHCI, pH 7.5, 10 *mM* $MgC1_2$, 1 *mM* DTT, 100 gg/mL BSA, 300 μM dNTPs, and reactions should be performed at 37°C. The enzyme is stable for over 6 mo when stored at -20°C in 25 *mM* Tris-HCI, pH 7.5, 250 *mM* NaCl, 5 *mM* DTT, and 50% glycerol.

TAQ POLYMERASE

Taq polymerase is a DNA polymerase isolated from the bacterium *Thermus aquaticus*. Similar enzymes have been isolated from other thermophilic organisms, including *Thermus thermophilus, Th. flavus, Th. litoralis, Pyrococcus furiosus,* and *Bacillus stearothermophilus*. *Taq* polymerase preparations have been isolated from several different strains of *Th. aquaticus* itself, and each has slightly different characteristics. The most common source is still strain YT1, although the enzyme is also available from recombinant clones

Enzyme Data

Taq polymerase is a monomeric enzyme with a mol mass in the range 60,000-97,000 Da depending on the source. The enzyme exhibits a 5'-3' polymerase activity with a high processivity and lacks a 3'-5'exonuclease (proofreading) activity, but does possess a 5'-3' exonuclease (strand replacement) activity. A modified form of the enzyme is now available from which this latter activity has been removed. DNA polymerases *from Thermus litoralis* and *Pyrococcusfuriosus* do exhibit the 3'-5' exonuclease activity, which renders DNA replication less error-prone.

Taq polymerase has another activity in that it can incorporate an additional nucleotide (usually deoxyadenosine) at the 3' terminus of any DNA duplex in a nontemplate-dependent reaction. This addition often hinders subsequent manipulations, but can be turned to advantage as a means of increasing cloning efficiency.

The major distinguishing feature of *Taq* polymerase is its extreme thermal stability. The enzyme can withstand temperatures in excess of 95°C for prolonged periods, and in fact, its optimum for reaction is 75°C. The rate of reaction is reduced to 50% at 60°C, and to 10% at 37°C. Arange of DNA poly merases from other species of thermostable bacteria (including those mentioned earlier) are now becoming available, which will broaden the spectrum of possible reaction conditions.

Uses of *Taq* Polymerase

Taq polymerase is rapidly replacing other DNA polymerases in many laboratory techniques, primarily because of its thermal stability and tolerance of temperature changes. This feature allows reactions to be performed at elevated temperatures, whereby DNA duplex melting and annealing can be rigorously controlled. This ensures the presence of a suitable single-stranded template for high-fidelity replication, and also the stringency of primer association can be precisely defined. Additional benefits are that contaminating enzyme activities are largely eradicated, and complex but routine procedures can be automated.

The major use of *Taq* polymerase at present is in the polymerase chain reaction (PCR). This technique is a simple method of amplifying minute quantities of DNA for a variety of subsequent procedures, including cloning, sequencing, hybridization, and genome mapping. The enzyme from *Thermus thermophilus* (when used in a manganese-containing buffer) has an additional reverse transcriptase activity, which extends the use of PCR directly to cDNA synthesis. The PCR has applications beyond the reseach laboratory, including uses in forensic science, disease diagnosis or prognosis, paternity testing, and in animal and plant breeding programs.

Another use of *Taq* polymerase is directly in DNA sequencing, where the high temperatures employed help reduce problems caused by secondary structure in the template and allow an increase in the stringency of primers used. *Taq* polymerase has, for the reasons mentioned earlier, become a very important tool for the molecular biologist.

ENZYMOLOGY OF THE *TAQ* POLYMERASE

Taq polymerase constitutes only 0.01-0.02% *of* the total protein of *Thermus aquaticus*. The increased demand for the enzyme led to the cloning of the gene in *E. coli*. An advantage of cloning an enzyme from a thermophile organism in a mesophile bacterium is that, in general, contaminants in the enzyme preparation should not be active at a temperature optimal for the *Taq* polymerase.

The sequence for the enzyme allowed a better characterization of the *Taq* polymerase. Given the biochemical, biophysical, and sequence data, it can be concluded that the two enzyme preparations published in the original paper represented either two different enzymes or that the smaller protein was a fragment of the constitutive *Taq polymerase*. The characteristic features of the commercially available cloned *Taq* polymerase are analogous to those of the latter protein.

The *Taq* polymerase gene is 2499 nucleotides long and codes for a protein of 832 amino acids. The theoretical mass is 93,900 Da, which is in fairly good agreement with the apparent mass of 97,300 Da obtained from gel electrophoresis. A sequence comparison with polymerase I from *E. coli* shows significant similarities in the 5'-3' exonuclease domain (amino acids) as well as in the polymerase domain (amino acids). Regions to which definite functions could be assigned in polymerase I are highly conserved in *Taq* polymerase.

No proofreading activity (3'-5' exonuclease) could be identified in *Taq* polymerase. There is no homology in *Taq* polymerase to a domain ranging from amino acid 300 to 410 in polymerase I, to which proofreading activity is attributed in the *E. coli* enzyme.

Taq polymerase catalyzes the DNA-dependent polymerization of dNTPs. One unit of the enzyme is defined as the amount of enzyme that will incorporate 10 nmol of radioactively labeled dTTP into acidinsoluble material at 80°C in 30 min. The enzyme could be purified to a specific activity of 200,000 U/mg protein. As an optimal reaction temperature, 75-80° C was established, at which the incorporation rate is approx 150 nucleotides/s/enzyme molecule. In the M 13 system, incorporation rates of <60, 24, 1.25, and 0.25 nucleotides/s/enzyme molecule were measured at 70, 55, 37, and 22°C respectively. Incorporation rates decrease at temperatures >90°C; the stability of the enzyme also decreases drastically. The half-life time of the enzyme declines from 130 to 40 to between 5 and 6 min at 92.5, 95, and 97.5°C respectively.

Enzyme activity is dependent on bivalent cations. Concentrations of 2 mM $MgCl_2$ are optimal. Maximum polymerization rates are obtained with 0.7-0.8 mM dNTPs. Substrate inhibition is observed at dNTP concentrations of 4-6 mM. Monovalent cations also have an effect on the activity of the enzyme. Optimum conditions are 50 mM KCl, whereas inhibition is reported at concentrations >75 mM KCl. NaCl, NH_4Cl, and NH_4 acetate cannot substitute for KCl without a decrease in specific activity. Denaturating agents, detergents, and organic solvents in low concentration are tolerated by *Taq* polymerase; at higher concentrations, inhibition is observed.

Table 11.2 : Inhibitor Effects on *Taq* Polymerase I Activity

Inhibitor	Concentration	% Activity
Urea	<0.5M	100
	10M	118
	1.5M	107
	2.0M	82
SDS	0.001%	105
	0.01%	10
	0.1%	<0
Ethanol	<3%	100
	10%	110
DMSO	<1%	100
	10%	53
	20%	110
DMF	<5%	100
	10%	82
	20%	17

A compilation of the effects of several substances is shown in Table elsewhere in this chapter. These concentrations are not obligatorily applicable to the PCR; e.g., 0.5M Urea are inhibitory to the PCR, whereas up to 1.5M Urea do not alter the incorporation yields of *Taq* polymerase. The inhibition by 0.01 % SDS can be reversed by adding 0.1 % Tween 20 or NP40.

Taq polymerase is an enzyme with high processivity. An absolute requirement is a DNA template with a primer to initiate the reaction. The primer must be dephosphorylated on the 3' end. *Taq* polymerase has a strand displacement 5'-3' exonuclease activity. 5'-phosphorylated oligodeoxynucleotides are not degraded by the 5'-3' exonuclease activity of the enzyme. Studies of the fidelity of DNAsynthesis by *Taq* polymerase revealed error rates of 1.1×10^{1} for misincorporation and 2.5×10^{-5} for frameshift mutations. An error rate of 2×10^{1} was determined by PCR experiments followed by sequencing and by denaturing gradient gel electrophoresis. T -C transitions are occurring more frequently than other misincorporation errors.

POLYMERASE CHAIN REACTION (PCR)

The first publication concerning the polymerase chain reaction (PCR) appeared in 1985. In this protocol, Klenow polymerase was used as the replicating enzyme, thus necessitating the addition of new polymerase after each denaturation step. In 1988, the first use of *Taq polymerase* for the PCR was reported, leading the way to an automation of the protocol. The application of *Taq* polymerase to the PCR was the basis for the success of the technique. The enzyme is extremely suitable for the following reasons:

- *Taq* polymerase is stable up to 95°C; thus it is not necessary to replenish the enzyme after each PCR cycle;
- The maximal enzyme activity is between the temperature range of 7075°C, which minimizes secondary structures of the template, resulting in high polymerization yield; and
- The annealing temperature can be chosen from 30-70°C, allowing an optimal adaptation of cycle parameters to appropriate annealing temperatures of the primers; therefore, byproducts are hardly generated.

Fundamentals of the PCR

The PCR is an enzymatic chain reaction that leads to the amplification of specific DNA sequences from a given template. In addition to the template, two primers (typically: oligodeoxynucleotides of 20 nucleotides in length) are needed, whose sequence must be complementary to the template DNA. The amplification is performed in the presence of dNTPs as substrate and catalyzed by *Taq* polymerase. The sequence between the two regions complementary to the primers is doubled by repeating the cycle, which consists of a series of temperature steps in which the DNA is denatured, the primers are

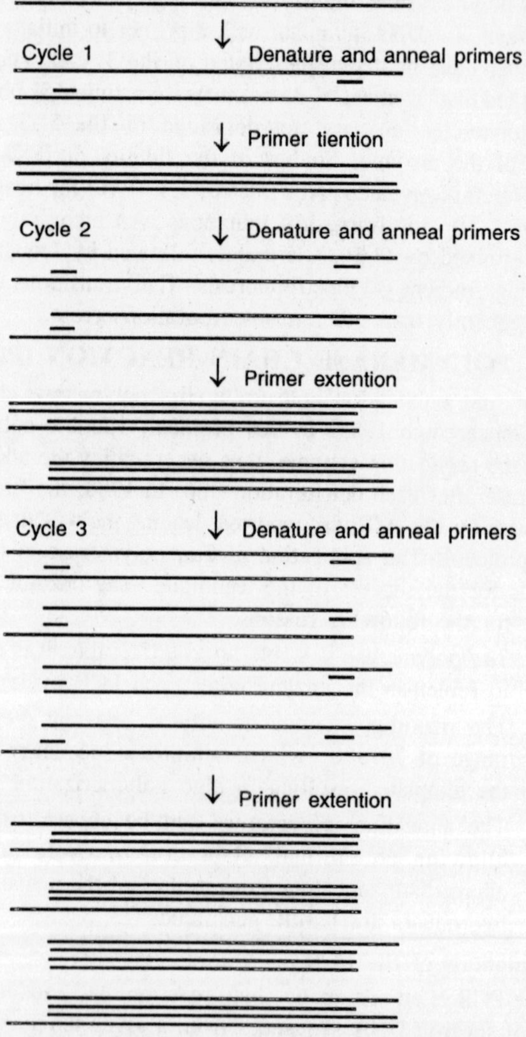

Figure 11.2 : Prmclple of the polymerase chain reaction.

annealed, and the polymerization of the specific sequence occurs. The newly synthesized sequence is available as template in the next cycle; therefore, theoretically, a twofold increase of template is achieved in each step. In theory, 30 cycles should yield an amplification of a specific sequence by a factor of $2^{30} = 1 \times 10^9$, as long as substrates or enzyme is not limiting.

The PCR technique reaches outstanding results because of the fact that the chain reaction requires the specific annealing of two primers to a complementary sequence. Unspecific annealing of primers or nicks in the template does not lead to exponential, but only linear amplification. Therefore, this system offers an optimal ratio between specific and unspecific signal.

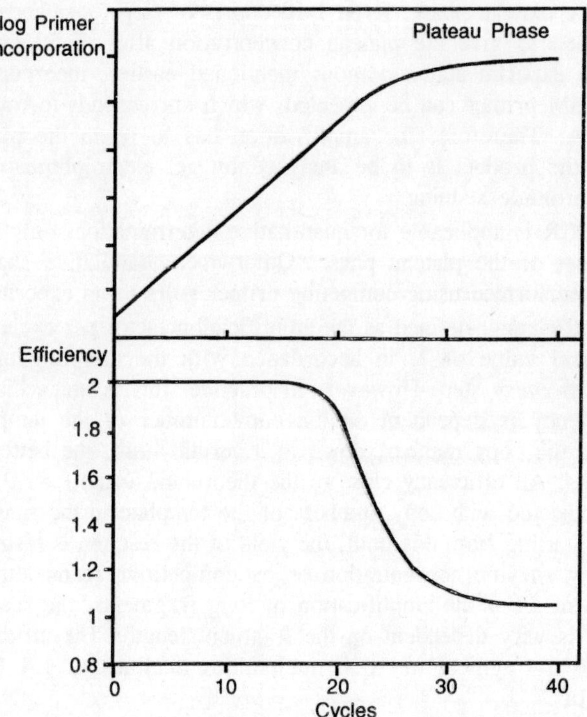

Figure 11.3 : Typical PCR Changes in efficiency and product yield Graphs were calculated on the basis of 1 μM of each primer spanning a stretch of 300 bp, 200 μM each dNTP, 2 U enzyme 1 nM (processing 10 templates/cycle, estimated enzyme degradation 5%/cycle), and 200-ng template (0 05 aMol single copy gene) Volume of 100μL.

Characteristically, < 1 μg of template in a vol of 100 pL is used in the PCR. In this experimental outline, a single copy gene of the human genome would be in a concentration of 2.5 fM. Primers and enzyme are used in large excess (1 μM and 1 nM, respectively). The efficiency of this reaction will be close to the theoretical value of 2, which means that the sequence is doubled in every cycle. After 20-30 cycles are completed, the product (and therefore the template) is amplified by a factor of 10^6. Now the template concentration is in the same order as the enzyme concentration. The efficiency of the PCR decreases steadily, and exponential growth is no longer observed. This phase of the reaction is called plateau phase. Hence, the product yield is dependent on the enzyme concentration.

The initial template concentration hardly influences the product yield in the plateau phase. Even 1-10 copies of target sequence will be amplified to give the plateau concentration after 40-50 cycles. Under the experimental conditions mentioned earlier, incorporation of 10100 nM primer can be expected, which corresponds to roughly 1 μg DNA. Therefore, the amplification has to reach the plateau phase, if the product is to be analyzed by gel electrophoresis and ethidium bromide staining.

The PCR is applicable for quantitative determinations only prior to the onset of the plateau phase. Otherwise, quantitative analysis has to be performed using competing primer pairs in an experiment.

The efficiency, defined as the amplification factor per cycle, has a theoretical value of 2, in accordance with the doubling of the sequence in every step. However, in practice, this is not achieved. The efficiency is dependent on the copy number of the template: The lower the copy number, down to a certain limit, the better the PCR works. An efficiency close to the theoretical value (>1.9) can only be expected with copy numbers of the template in the range of 10^7 -10^8. Starting from this limit, the yield of the reaction is restricted by too low enzyme concentration or by competitive reannealing of the fragments. For the amplification of long fragments, the reaction efficiency is very dependent on the fragment length: The efficiency of 1.9 for a fragment of 1000 nucleotides declines to 1.4 for a fragment of 7 kb.

Application of Taq Polymerase in the PCR

The application of *Taq* polymerase in the PCR requires certain prerequisites to be fulfilled.

Buffers

The buffer used most commonly in the PCR consists of 10 mM

Tris-HCI, pH 8.4, 50 mM KCI, 1.5-2.5 mM $MgC1_2$, and 0.01 % gelatin as stabilizer. The Mg^{2+} ion concentration influences predominantly the specificity of the PCR. With increasing Mg^{2+} concentrations, the amplification of unspecific sequences increases. At high dNTP concentrations, it should also be taken into account that the concentration of free ions is reduced as a result of Mg^{2+} binding to dNTPs and to DNA. It was reported that, in an incorporation experiment, high dNTP concentrations (4-6 mM) were inhibitory.

A different widely used buffer system is composed of 67 *mM* Tris HCI, pH 8.8, 6.7 mM $MgC1_2$, 16.6 mM $[NH_4]_2SO_4$, 10 mM (3-mercaptoethanol, and 0.02% bovine serum albumin (BSA) or gelatine. In principle, gelatine should be given preference over BSA, because BSA is likely to denature. On the other hand, in some protocols the use of stabilizing proteins is totally abandoned. Sometimes 10% dimethylsulfoxide (DMSO) is added to the buffers mentioned above.

The Michaelis Menten constant K_m of *Taq* polymerase for each dNTP has a value of 10-15 *μM*. A normal PCR experiment contains 50-200 *μM* of each dNTP. This concentration is sufficient to amplify even large fragments to microgram yields without any limitation of dNTP substrate. Even after 60 cycles, a pool of about 50% dNTPs is still available. Sometimes the dNTP concentration is lowered to 10 pM in order to achieve an optimum incorporation of radioactive labeled deoxynucleotides. Modified nucleotides, e.g., deoxynucleoside-a-thiotriphosphates (dNTPαS), 7-deaza-2'deoxyguanosine-5'-triphosphate d(c^7GTP), 5-Biotin (11)-2' deoxyuridine-5'- triphosphate (Bio 1 1 dUTP), are incorporated at high rates by *Taq* polymerase.

Primer

In theory, at least a 16 by of primer are necessary for an unambiguous hybridization to a given stretch of DNA in the human genome (3×10^9 bp). In practice, primers 20-25 nucleotides long are used in PCR experiments, with the G-C content varying from 40-60% in order to ensure specific amplification. Primers shorter than 20 nucleotides are used successfully for allele-specific hybridization. Except for this purpose, utilization of shorter primers should be avoided, since they might detach from the template at the high temperature where polymerization is effective.

To determine the approximate melting temperatures of a primerDNA duplex, the equation:

$$T_m\,[°C] = 2[T + A] + 4[G + C]$$

can be used. More exact numbers can be calculated by applying algorithms that take the occurrence of intramolecular interactions, e.g., hairpin structures, into account.

In established oligodeoxynucleotide synthesis, a product yield of more than 80% for short primers can be expected. Therefore, a purification of the crude synthesis products is most often not required. Depending on circumstances, which demand purification of the crude products, e.g., an oligodeoxynucleotide has to be labeled prior to the PCR, two methods are suitable for purification: When little amounts of primers are needed, denaturing PAGE with *7M* urea followed by detection by shadow casting is a convenient method for purification. Several oligodeoxynucleotides can be purified in parallel. Since urea is an inhibitor of *Taq* polymerise, it must be removed by chromatographic techniques after elution from the gel slice. Chromatographic procedures are a better choice for the purifications of large quantities of oligodeoxynucleotides. Reversed-phase HPLC techniques and ion-exchange chromatography are commonly used.

Primers should always be stored lyophilized in aliquots at -20°C. It is convenient to prepare l0X stock solutions with 1-10 μM concentrations of the primer. A primer concentration of 0.1-1 μM is optimal for the PCR. An increase in the oligodeoxynucleotide concentration will produce a rise in nonspecific bands. One example of the production of an unspecific band is owing to partial primer duplex formation and polymerization and leads to a product of less than the length of the two primers.

Apparently, primer pairs are formed and elongated by the enzyme. Biased primer concentration results in asymmetric amplification of one strand.

PCR is capable of dealing with many different problems by utilization of primers specifically designed for a particular purpose. Primers carrying a chemical label at the 5' end, e.g., oligodeoxynucleotides substituted with a biotin or fluorescent dye, are well accepted by *Taq* polymerase. Oligodeoxynucleotides can be synthesized with a primary amino group (aminolink) at the 5' end, to which biotin or fluorescent markers may be coupled. As a consequence, 5' biotinylated primers offer the option to exploit avidin-coupled dye reactions as well as the immobilization of PCR products via an avidin or streptavidin bridge to a matrix. The use of fluorescent primers offers a variety of applications in qualitative and quantitative analysis, e.g., nonisotopic sequencing reactions or color coupling tests.

A perfect homology to the target sequence is not a *conditio sine qua non* for primers, although a high degree of sequence homology is necessary. Taking into account that the enzyme is working downstream from the 3' end of the primer, only the last three bases adjacent to this region need to be correctly base paired, in order to initiate polymerization. The 5' region of the primer is less sensitive to base mismatches. Therefore, new restriction sites can easily be introduced into an amplification product. It is also possible to use primers elongated by a stretch of 40 GC by. Successful amplification of sequences derived from known amino acid sequences utilizing primers with mixed bases at one position has been reported.

Template

Normally, very much less than 1 μg DNA is used for an experiment. The sensitivity of the method allows a dilution of the sample down to the theoretical limit of 1 molecule/test, even for routine diagnostics. The requirements for DNA purity are rather modest. Intact cells may be used; they are lysed during the first denaturation step of the PCR procedure. Some authors recommend a proteinase K digest in the presence of nonionic detergents in order to render the DNA more accessible. However, there is no advantage to employing higher concentrations of template; this will make a contamination with inhibitors more likely. Without any DNA isolation, amplification of bacterial sequences is feasible directly by incubating 1×10^3 cells at 95°C for 10 min. At higher cell concentrations, the PCR is inhibited. Hematin (hemin) has also been found to inhibit the PCR at levels as low as 0.8 *μM*. Therefore, only nuclear fractions from whole blood samples should be analyzed by the PCR. The sensitivity of the technique allows the amplification of degraded DNA from very old, but well-preserved samples. A preceding reverse transcription in the same buffer system permits the direct amplification of RNA sequences.

Taq Polymerase

Taq polymerase 1 is commercially available from several manufacturers at a concentration of 5 U/μL. One unit corresponds to 50 fmol of enzyme. *Taq* polymerase lots from different suppliers may lead to considerably different quantitative results. One to four units per 100-μL reaction aliquot are sufficient to amplify a sequence to yield amounts of DNA detectable by ethidium bromide staining. Larger enzyme quantities will provoke the amplification of unspecific products. If the denaturation temperature for double-stranded DNA is optimal, enzyme damage can be minimized, and enzyme activity will be present in the reaction mixture even after 50 or more cycles. The enzyme dilution should always be made up freshly by pipeting from stock solution into incubation buffer.

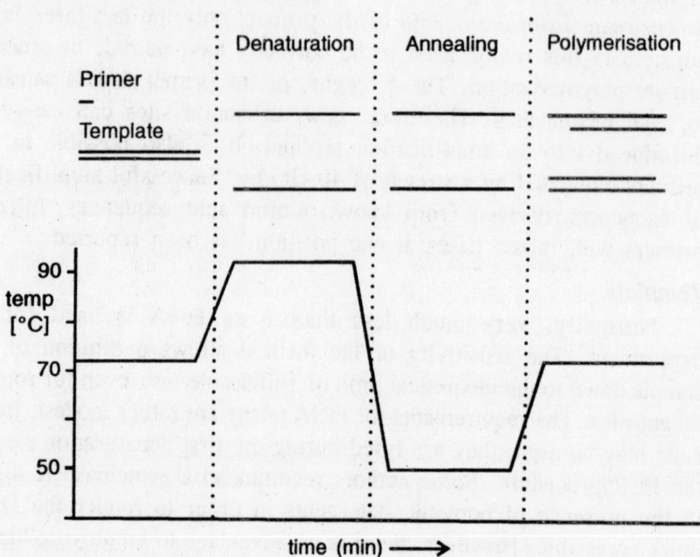

Figure 11.4 : Typical PCR temperature protocol.

Thermocycler

The PCR reaction requires a repeated cycling through temperature profiles. Sample volumes of 20-100 μL are used in small reaction tubes, which will facilitate the heat transfer from the incubation device. Temperature control is usually achieved by automated thermocyclers:

- Water bath devices;
- Heating blocks with conventional electrical heating and liquid cooling systems; or
- Heating based on other technologies, e.g., Peltier elements or fan heating.

It is the main feature of the devices to generate fast temperature shifts in the range of 30-95°C. Deviations concerning block uniformity or precision in repeated cycles of more than 1°C are not tolerable. A prerequisite for allele-specific amplification experiments is a highly reproducible annealing temperature.

With regard to the adaptation of published protocols to a specific thermocycler, it is very important to consider the way the actual temperature is determined and regulated. Three principles can be distinguished:

- Temperature assessment in the heating block: (a) the time reading starts, when the desired value in the block is reached or (b) the time reading starts independently of the effective temperature of the block;
- Temperature assessment in a reference vial: the time reading starts, when the rated temperature in the reference is reached; and
- Temperature assessment in water baths: the time reading is independent of the actual sample temperature.

Different from the programmed temperature profile of the heating block, the actual profile in the reaction vial is distinctly more shallow, since the heat transfer through the material of the vessel is fairly bad. For a successful PCR experiment, a set of defined temperatures must be adjusted:

- To denature the template, 90-95°C must be reached for a short interval. Prolonged incubation times at this temperature will lead to severe enzyme losses by degradation. It was demonstrated that complete strand separation is not a necessary condition for annealing, rather, a partial melting of the strands is sufficient to permit hybridization of the primers.
- The annealing temperature must be attained for a few seconds only. At a primer concentration of I μM, the half-life time of the templateprimer hybrid is not more than 0.4 s at a temperature 20°C below the T_m value (44).
- As mentioned earlier, the incorporation rate of the enzyme is 60-150 nucleotides/s at 70-75°C; thus, a few seconds are adequate for the amplification of sequences shorter than 1000 nucleotides. An incubation time of 1 min/1000 nucleotides of sequence length is a good rule of thumb, when sequences exceeding a length of 1000 nucleotides or more are to be amplified.

Standard PCR Protocol

Following this protocol, a product band should be detectable from a single copy gene. Prepare the following solution in a final volume of 30-100 μL:

- 1 μg Genomic DNA as template;
- 1 μM Primers, 20-25 by in length, flanking a sequence of 100-1000 bp;
- 10 mM Tris-HCI, pH 8.4;

- 50 mM KCl;
- 1.5 mM $MgC1_2$;
- 0.01% Gelatine; and
- 200 μM of each dNTP

Top this mixture carefully with paraffin oil, incubate at 95°C for 3 min, add 1-4 U of *Taq* polymerase/100 μL reaction vol, and repeat 30 cycles (temperature measurement in the heating block) of:

- 1 min at 74°C;
- 1 min at 92°C; and
- 1 min at 50°C.

End the procedure with an incubation at 74°C for 3 min. Analyze a 10-μL aliquot on an agarose or PAGE gel; stain with ethidium bromide.

Reliable PCR Experiments

In clinical diagnosis, reliability is the most important criterion for the application of a test. The extreme sensitivity of the PCR requires high standards for the conditions of operation. Even minor contaminations with foreign DNA will render results uncertain. With this caveat in mind, some precautions are necessary to avert false positive results.

- Separate all PCR materials and equipment from processes involved in DNA isolation.
- Aliquot all buffer, dNTP, and primer stock solutions.
- Pipet the DNA last.
- Always process controls in parallel: (a) a negative control with all components, but lacking the template, and (b) a positive control with a control template in a concentration similar to the template under investigation.
- Perform only the minimum number of amplification cycles necessary to detect the product. This is crucial for experiments in which the loss of a band (e.g., demonstration of the deletion of particular a sequence) has to be detected.

Recently, it was reported that irradiation at 254 nm for 5 min of the PCR mix lacking template and enzyme will reduce DNA contamination by a factor of 10^5.

PCR APPLICATIONS

PCR as a Standard Method in Clinical Diagnosis

The scope of PCR application in clinical diagnosis ranges from prenatal diagnosis for sickle cell anemia to postmortem analysis of tissues of cancer metastases. We would like to restrict this chapter to some examples with characteristic features of PCR application in clinical diagnosis.

The reasons for the success of the PCR as a standard method in clinical diagnosis are its outstanding sensitivity, reliability, and versatility. Diagnostic screening for mutations by the PCR should always be complemented with a statistical analysis; the probability of false diagnosis is on the order of 1% for a single-locus analysis in single cells.

The extreme sensitivity of the technique can trace one target sequence in a background of 10^5-10^6 cells. The PCR makes affirmative statements on the genetic information of single somatic or sperm cells possible. In a serial examination, 700 different sperms were typified, demonstrating a frequency of recombination of 0.16 in one individual. An extension of this test series could allow the detection of recombination frequencies <0.001; hence, data would be available that are more exact than information from linkage analysis of pedigree studies.

Multiple gene loci can be amplified in parallel. Six domains of the Duchenne Muscle Dystrophy (DMD) gene were amplified in a single experiment. Thus, it is possible to monitor by PCR 70% of the deletions in the 700-kb DMD gene. A simple test for β-thalassemia was developed by combining the PCR amplification with subsequent dot blot hybridization to allele specific oligodeoxynucleotides.

The PCR was established to amplify fragments up to 3000 by to amounts detectable by staining. The amplification of larger fragments is feasible, although with little efficiency: Hypervariable minisatellite alleles up to 5-10 kb were faithfully amplified. Up to several micrograms of product can be achieved by the PCR, amounts sufficient as substrate in binding experiments, e.g., footprint analysis.

PCR Applications in Combination with Nonradioactive Detection Techniques

The use of radioactive probes should be avoided in routine screening tests. Even DNA of low abundance can be analyzed by the PCR followed by restriction analysis of the products. The definite sequence is enhanced to high copy numbers; therefore, radioactive detection systems are dispensable. Product identification can be easily done, for example, by hybridization to biotinylated oligodeoxy nucleotides and subsequent detection via avidin-linked enzymes.

Allele-specific oligodeoxynucleotide (ASO) PCR is based on the fact that only fully complementary primers will initiate the reaction. Therefore, amplification will be prohibited, if the complementary sequence is deleted or mutated. A primer specific for the mutated sequence is used as positive control. Point mutations, deletions, and insertions in the a-l-antitrypsin gene were monitored with primers harboring a 3' base mismatch with respect to the unmutated gene. One 3' mismatch is necessary and sufficient to ensure reliable detection. Primers bearing a mismatch located in the center of the primer sequence are not suitable for allele-specific hybridization.

A different technique to find point mutations is competitive oligodeoxynucleotide priming *(COP)*. When the annealing of a primer mixture is performed at low stringency, the perfectly matched primer will be favored with a level of discrimination greater than 100:1. An extension of COP and ASO-PCR involves the use of oligodeoxynucleotides conjugated with a fluorescent dye marker; the products are analyzed in a fluorimeter, hence rendering the electrophoretic analysis obsolete.

Application of Specifically Designed Primers

For the reason that a primer needs to be complementary only at its 3' end to the sequence to be analyzed, mismatched primers can be of advantage in introducing new restriction sites in the amplified product. The product can be traced by restriction enzyme analysis even at high background. Also, the fragment produced can be cloned readily, one more benefit of this method.

Sometimes it is difficult to detect a single mutation in longer fragments using denaturation gradient gel electrophoresis (DGGE). This drawback can be overcome by attachment of a 40-bp GC-rich stretch to the primers. The analysis of the PCR product, melting at a higher temperature, will enhance the sensitivity of this method.

Mixed oligodeoxynucleotide primers designed on the basis of the amino acid sequence of urate oxidase were used successfully to clone the gene for this enzyme. If the primers crosshybridize to repetitive sequences, even an optimal adaptation of the annealing temperature will not eliminate unspecific bands. In this case, the amplification products can be diluted and reamplified with a second set of primers. The use of dc^7GTP in a ratio of 3:1 over GTP is recommended to improve the PCR of sequences with distinct secondary structures.

Normally, it is a prerequisite for the PCR to know the sequences flanking the region to be amplified. Different methods were developed to circumvent this drawback: Circularization of DNA fragments and subsequent cleavage in one known site result in a constellation, where an unknown region is flanked by sequenced regions.

Normally, the PCR is applied to generate well-defined sequences. Whole genome PCR involves the ligation of oligodeoxynucleotides prior to the amplification, and constitutes a pool of material for further manipulation. With this technique, possible pitfalls in microcloning of dissected chromosome fragments could also be bypassed. After the ligation of oligodeoxynucleotides to restriction fragments, the DNA was amplified and cloned in high yields.

Ligation of known sequences to chemically cleaved DNA templates and subsequent PCR amplification allows a specific in vivo footprint analysis as well as detection of methylation patterns. The principles of this ligation-mediated PCR application are depicted in Figure elsewhere in this chapter. Random cleavage sites are generated in genomic DNA using Maxam-Gilbert chemistry. The strands are denatured, and a genespecific primer is annealed and extended with DNA polymerase. Oligodeoxynucleotide linkers are blunt-end ligated to the doublestranded products. The linker-ligated fragments are amplified with the PCR and analyzed on sequencing gels.

AMPLIFICATION OF mRNA

The amplification of RNA is performed on a cDNA template produced by reverse transcriptase. A small and not very efficient reverse transcriptase activity was detected in *Taq* polymerase. Molony Murine Leukemia Virus (MMLV) reverse transcriptase requires buffer conditions similar to *Taq* polymerase. Therefore, after addition of poly dT or sequence-specific primers and reverse transcriptase, the PCR can be executed directly in the same vial. Avian Myeloblastosis Virus (AMV) reverse transcriptase and *Taq* polymerase may also be added together to the RNA; the reactions will be started subsequently by choosing the proper reaction temperatures.

Starting with cDNA, fragments as long as 1.6 kb could be cloned. The extreme sensitivity of the PCR allows semiquantitative recordings on specific mRNA content of single cells. Eleven out of 12 patients suffering from chronic myelogenous leukemia (CML) who underwent bone marrow transplantation still produced transcripts specific for CML.

Although the patients were cytogenetically inconspicious even 3-6 yr after surgery, CML-specific transcripts could be detected by PCR of the reverse transcribed RNA.

If the boundaries of genes are not exactly known, ligation of

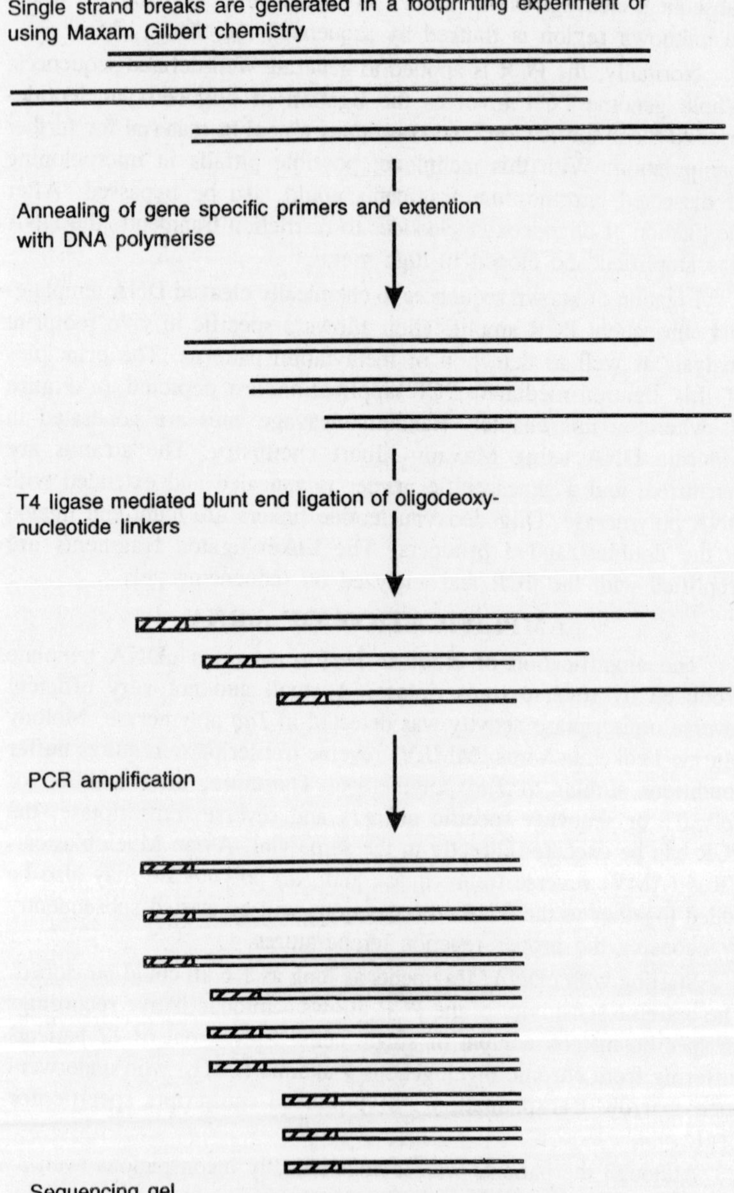

Figure 11.5 : Application of PCR to in vivo footprinting and detection of methylation pattern.

specific sequences to the cDNA will render an amplification feasible. Primers complementary to the ligated oligodeoxynucleotides permit the PCR of unknown sequences. This method is termed "anchor PCR". An alternative to cDNA synthesis is a combination of consecutive RNA-DNA hybridization, S 1 nuclease digest, and PCR. This technique was applied successfully to trace CML-specific transcripts.

Sequencing of PCR Products

Sequencing PCR-amplified products is a time-saving procedure in the analysis of genes. No time-consuming cloning of DNA is necessary. On the contrary, errors made by *Taq* polymerase will lead to false sequence information, if cloned PCR products are sequenced. Sequencing revealed that, in 22 clones created from amplified DNA, none had a faultless sequence. In a direct sequencing reaction of amplified DNA, on the average, these shortcomings will not become visible. Double-strand sequencing of a single copy gene was utilized to typify mutations in the globin gene.

In order to increase the efficiency of the sequencing reaction, diverse strategies were developed aimed at yielding single-stranded DNA. When biased primer concentrations are used, a phase of normal PCR will precede a reaction producing dominantly single-stranded DNA. A well-balanced ratio of the primer concentration is crucial. If the primer concentration is too low, too little DNA will be available for single-strand synthesis. On the other hand, the ratio of single-stranded to double-stranded product will be inferior at primer concentrations that are too high. In experiments working with asymmetric primer concentrations, the PCR efficiency decreases to a factor of 1.7 as compared to amplifications with stoichiometrlc concentrations.

Single-stranded DNA can be recovered by affinity chromatographic purification after amplification with biotinylated primers. In an elegant experiment, a phage promoter was attached 5' onto at least one of the PCR primers. The amplified segments were transcribed into RNA and sequenced with reverse transcriptase. Another way to generate single-stranded DNA involves a 5'-phosphorylated primer and phage X exonuclease, which specifically attacks double-stranded DNA only if there is a 5' terminal phosphate. With the dideoxy sequencing method, single copy genes can easily be sequenced with fluorescent dye-labeled primers. As an alternative, Maxam-Gilbert sequencing of PCR products or sequencing after incorporation of dNTPaS during the PCR is possible. When fluorescent dye-labeled primers are utilized, the cleavage products can be sequenced directly.

In the plateau phase of the reaction, the processivity of *Taq* polymerase decreases at excess template concentrations. This phenomenon is exploited in a very simple DNA sequencing method, where phosphothioates are incorporated into the newly synthesized DNA. The product is digested with snake venom phosphodiesterase and analyzed on a sequencing gel. No product purification is necessary applying this technique.

PCR AND SITE-DIRECTED MUTAGENESIS

High yields of the desired mutations can be produced, when mismatched primers are employed in the PCR. Deletions, insertions, and mutations were created in plasmids with the support of mismatched primers and PCR. Furthermore, these primers may generate a new restriction site, serving as an option to introduce amplified fragments with the cassette method.

Amplification of the whole plasmid offers the opportunity to introduce mutations in the absence of restriction sites; partial deletions in the plasmid are also feasible. In 70% of the amplified plasmids, 3' additions of one base not present in the template were observed. Prior to cloning, these protruding ends had to be removed by the Klenow fragment of DNA polymerase to restore the gene sequence. The PCR has proved useful in the precise gene fusion at any chosen location. Sequencing of mutant clones after mutagenesis is compulsory, given the relatively high error rate of *Taq* polymerase (1.1×10^{-4}, 2×10^{1}). At least 20% of the clones derived from a 250-bp fragment amplified 10^6-fold will contain one error in the sequence.

QUANTITATIVE INTERPRETATION OF PCR RESULTS

Standard PCR in itself is not a quantitative method, when the plateau is reached. For a quantitative interpretation of PCR experiments, low cycle numbers and template concentrations below enzyme concentrations are optimal, i.e., conditions where amplification is not restricted by limiting parameters prevailing in the plateau phase. Hybridization of ^{32}P-labeled oligodeoxynucleotides revealed a linear correlation between template DNA and detected signal, provided that 500-4000 copies of DNA as template and 30 amplification cycles were used. In order to increase the incorporation rate, the dNTP concentration was lowered to 10 μM.

In a technique referred to as differential PCR, alterations of genes

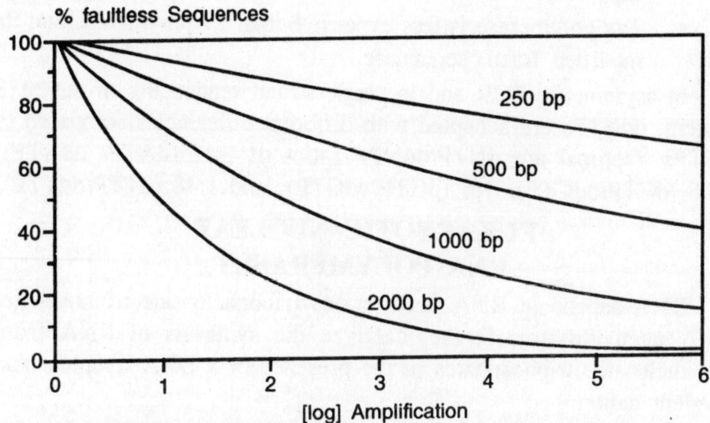

Figure 11.6 : Dependence of error rate and fragment length.

can be monitored when two genes are coamplified in one reaction vessel. In this method, the primers are competing for the limiting enzyme; therefore, gene copy numbers can be estimated semiquantitatively even in the plateau phase of the PCR.

Quantitation of mRNA transcription in different cell types was successfully performed using Southern hybridization to an internal standard and to the amplified product. Monitoring antiviral therapy benefits from the exceptional sensitivity of the PCR as well.

A test system where standards identical to the target sequence, except for one restriction site, are coamplified provides an elegant way for the absolute determination of mRNA amounts. Consequently, both signals are enhanced to the same extent; variations in the reaction efficiency will not influence the result. After completion of the PCR, the standard is cleaved and compared with the target sequence. An alternative to blotting techniques is the utilization of biotinylated primers; the PCR products are immobilized on avidin agarose and hybridized to specific probes.

SEQUENCING WITH *TAQ* POLYMERASE

Although *Taq* polymerase is considered often to be only useful for the PCR, last but not least, it should be pointed out that the enzyme is ideal for dideoxy sequencing, the reasons being:

- The enzyme is highly processive;
- Secondary structures are suppressed at the optimal temperature range of 70-80°C;
- dc^7GTP is accepted as building block; this nucleotide analog

suppresses secondary structures as well; and

- *Taq* polymerase is less expensive than T7 polymerase and its modified form, sequenase.

In asymmetric PCR and in single-strand sequencing in the M13 system, ddNTPs are accepted with different efficiencies compared to dNTPs. Optimal are dNTP/ddNTP ratios of 1:32 (dATP/ ddATP), 1:16 (dCTP/ddCTP), 1:6 (dGTP/ddGTP), and 1:48 (dTTP/ddTTP).

EUKARYOTIC NUCLEAR RNA POLYMERASES

DNA-dependent RNA polymerases (ribonucleoside triphosphate: RNA nucleotidyltransferase) catalyze the synthesis of RNA from ribonucleoside triphosphates in the presence of a DNA template and divalent cation.

The RNA molecules are synthesized complementary and antiparallel to one of the DNA strands in a 5' to 3' direction. The ribonucleotides are covalently joined together by internucleoside 3' to 5' phosphodiester bonds with concomitant release of inorganic pyrophosphate.

In order for the polymerase to do this, it must "recognize" the beginning of the DNA sequence to be transcribed within the double-stranded template; insert the correct nucleotide residue into each position, as determined by the template sequence; carry out synthesis, so that RNA is synthesized from beginning to end as a consequence of one polymerase binding event; and recognize termination signals in order to end RNA synthesis at the appropriate point. To achieve all this, the polymerase must be able to recognize regulatory sites on the DNA template and interact with protein factors that modulate the activity of the enzyme. Several reviews have appeared on eukaryotic RNA polymerases.

BIOLOGICAL FUNCTION

Bacterial RNA polymerase containing a single apo enzyme is responsible for all the synthesis of RNA in the cell. There are approx 7000 RNA polymerase molecules present in an *Escherichia coli* cell, and many of these are involved in transcription, although the actual number depends on the growth conditions. The RNA polymerase best characterized is that of *E. coli,* but its structure is similar in all other bacteria studied. The holoenzyme has a mol mass of about 480,000 Da and is made up of five subunits, α, β, β', σ, and ω, there being two copies of α and one each of the others: $\alpha_2\beta\beta'\sigma\omega$. The holoenzyme can be separated into two components, the apo enzyme ($\alpha_2\beta\beta'\omega$) and the sigma factor ($\sigma$). The sigma subunit of the E. *coli* polymerase has a specific role as an initiation factor for transcription: It enables the enzyme to find consensus promoter sequences. Only the holoenzyme can initiate transcription; the sigma "factor" is then released, leaving the core enzyme to undertake elongation. Thus, the core enzyme has the ability to synthesize RNA on a DNA template; its affinity for DNA is the result of electrostatic attractions, but is nonspecific. It is the sigma factor that ensures correct transcription initiation involving the stable binding of the RNA polymerase to the promoter. Eukaryotes, however, contain multiple DNA-dependent RNA polymerases. Those that transcribe nuclear genomes are generally referred to as RNA polymerases I, II, and III transcribing rRNA, mRNA and tRNA, and 5SrRNA, respectively.

Properties

Eukaryotic RNA polymerases transcribe both double-stranded and single-stranded DNA templates in vitro, and single-stranded templates generally promote higher rates of transcription. RNA polymerases of plants and animals are capable of initiating transcription at nicks in the DNA template in vitro. Purified eukaryotic RNA polymerases are capable of some degree of selective binding and initiation of RNA chains on deproteinized DNA templates, although this is nonspecific, and in no case has the selectivity been correlated with a promoter or initiation site. From a number of studies with animal cell-free extracts, it is apparent that additional factors are required to obtain faithful transcription in vitro, so the enzymes are not often used in their pure state because they lack specificity. Eukaryotic RNApolymerases differ in subunit structure, type of genome transcribed, chromatographic properties, divalent cation, ionic strength optima, and α-amanitin sensitivity. The poisonous mushroom *Amanita phalloides* produces α-amanitin, a bicyclic octopeptide that blocks the elongation step of transcription by direct interaction with RNA polymerases that differ in their sensitivity. RNA polymerase I is insensitive to the toxin, but is responsive to actinomycin D, which binds to DNA preventing elongation. RNA polymerase II is very sensitive and RNA polymerase III moderately so. Thus, the various RNApolymerases I, II, and III are readily distinguished. Intact cells are not freely permeable to α-amanitin; therefore, it is necessary to use a whole-cell or nuclear extract.

Ion Requirement

The requirement for a divalent cation by RNA polymerases indicates binding of nucleotides to the enzyme as metal chelates, for example: MgATP. The optimum divalent cation concentration is generally in the range of 5-10 mM Mg^{2+} or 1-2 mM Mn^{2+}. In addition to the above requirements, RNA polymerase activity is stimulated by the inclusion of monovalent cations, such as $(NH_4)_2SO_4$, NH_4C1, KC1, or NaCl. This is especially evident when engaged RNA polymerases are assayed in isolated nuclei or chromatin. With isolated plant or animal nuclei, RNA polymerase I and III activity is most active at 0.05-0.1 OM $(NH_4)_2SO_4$, whereas RNA polymerase II activity is optimized at 0.250.50M.

Structure

The nuclear RNA polymerases have a complex subunit structure. Each has a mol mass of approx 500,000 Da and is made up of two large subunits, each having a mol mass >100,000 Da, which seem to be characteristic for each class of enzyme, and a number of smaller subunits, with mol wt from 10,000 Da to just less than 100,000 Da. The situation is further complicated by the phenomenon of microheterogeneity within RNA polymerase classes.

TRANSCRIPTION FACTORS

The basic components required for efficient and accurately regulated eukaryotic transcription initiation include two types of DNA elements known as promoters and upstream regulatory sequences; two sets of proteins known as general transcription factors and specific or regulatory proteins, and the RNA polymerase enzyme. Transcription can occur when the eukaryotic RNA polymerase recognizes a preexisting DNA-protein complex. RNA polymerase II has four general transcription factors associated with it; these have been partially purified and designated TFIIA, B, D, and E. TFIIA is required for the efficient interaction of TFIID with the promoter element known as the TATA box, which is found in nearly all eukaryotic promoters, with a consensus sequence "TATAAA" and located approx 25-30 by upstream of the transcription start site in mammalian promoters. It has been shown that TFIIE specifically interacts with RNA polymerase II and TFIIB. These transcription factors range in mol mass from 25,000 Da for yeast to 76,000 Da for TFIIE.

Although promoters themselves are sufficient to direct basal level gene expression, specific transcription requires the assistance of additional transcription factors. These sequence-specific DNA binding proteins are directly involved in the recognition of mRNA transcription initiation in higher organisms; for example, CAT transcription factor (CTF) is responsible for selective recognition of eukaryotic promoters that contain the sequence CCAAT, whereas the Sp 1 protein binding regions contain one or more GC boxes (GGGCGG), which can be present in either orientation with respect to transcription. Not all GC boxes bind Sp1 with equal affinity, and sequences outside the core hexanucleotide seem to modulate the efficiency of binding.

In HeLa cell factor, termed the major late transcription factor or upstream stimulatory factor, MLTF/USF binds to an upstream consensus sequence GGCCACGTGACC, which functions in a bidirectional manner. MLTF/USF may act synergistically with TFIID, since binding of MLTF/USF enhances binding of TFIID. The binding sequence occurs in a number of cellular promoters, for example, mouse and rat γ-fibrinogen. Another promoter-specific transcription factor is the heat shock transcription factor (HSTF). All major heat shock genes contain multiple HSE (heat shock regulatory elements) with the consensus sequence C-GAA-TCC—G. In HeLa cells, heat shock causes more HSTF to bind to the HSE, whereas in yeast, HSTF binds constitutively and heat shock causes HSTF to become phosphorylated.

In Vitro Transcription Systems

The development of methods for preparing soluble whole-cell and nuclear extracts from mammalian cells has led to a dramatic increase in our understanding of the mechanistic details of complex biochemical reactions, such as transcription, replication, and RNA processing.

In vitro transcription systems were partially developed to study the biochemical mechanisms of DNA-protein interaction, gene expression, and regulation. To date, several cell-free transcription systems have been developed, the first by Wu in 1978; which demonstrated accurate transcription of exogenous DNA by polymerase III, and others such as those developed by Weil et al.; Manley et al.; Dignam et al.; and Zerivitz and Akusjarvi.

The major advantage of soluble cell-free transcription systems is that they can be fractionated by standard techniques in order to identify those components involved in the transcription process. The purified enzymes alone do not direct selective and accurate transcription from purified DNA templates, but they appear to initiate and terminate at a large number of sites, which may not be relevant to transcription in vivo. The most extensive use of in vitro transcription systems to date has been to study the structure of promoter sites and the transcription factors needed to bring about selective transcription.

EXPERIMENTAL PROCEDURES

TCA Precipitation Assay

This procedure can be used as an estimate of nucleoside triphosphate incorporation into trichloroacetic acid (TCA) precipitable products following an in vitro transcription reaction. ^{3}H-UTP is usually the radiolabel of choice. The reaction (as detailed in Section 4.2., but for the substitution of ^{32}P-labeled GTP with 1.0 μCi of ^{3}H-UTP) is stopped by the addition of 10% TCA. Acid-insoluble material is then collected on glass fiber disks, washed twice with 5% TCA (with sodium pyrophosphate to reduce nonspecific binding of the radiolabel), and finally with alcohol to remove unincorporated radiolabel. The filters are dried and the radioactivity estimated by using a scintillation counter; hence, an estimation of nucleotide incorporation can be made.

Run-Off Transcription

RNA polymerase does not accurately terminate transcription in vitro. One procedure therefore is to use as template a cloned gene fragment that includes the promoter, but is truncated downstream. Using the "run-off" assay system, the whole-cell or nuclear lysate, DNA template, and radiolabeled (^{32}P) nucleotide are incubated at 30°C for 60 min. RNA is synthesized from the truncated DNA template and analyzed electrophoretically to detect the species of RNA with the predicted length.

A typical 50-μL in vitro transcription reaction would contain the following:

- 20-30 μL Whole-cell or nuclear extract;
- 1-2.5 μg Template DNA;
- 500 μM each of ATP, CTP, and UTP;
- 4 mM Creature phosphate;
- 50 μM Unlabeled nucleoside triphosphate (GTP)
- 10 μCi Labeled nucleoside triphosphate (^{32}P-GTP specific activity).

The volume is made up to 50 μL with sterile distilled water and incubated at 30°C for 1 h. Suitable controls include omission of template DNA or unlabeled nucleoside triphosphates, and addition of RNase or α-amanitin.

This system uses template DNA that has been cleaved by a

restriction enzyme cutting downstream from its putative start site. RNA polymerases that transcribe this DNA will stop or "fall off" when they reach the end of the DNA. If a substantial number of enzymes initiate transcription at the same site, then a population of molecules will migrate as a single band during gel electrophoresis. If DNA segments cleaved by different restriction enzymes are used as templates in separate reaction mixtures, the transcription start site can be deduced by comparison of the sizes of RNAs produced; this can then be used for promoter mapping.

Primer Extension

This method is commonly used to measure the precise 5' end of transcripts, based on the fact that reverse transcriptase can transcribe RNA into DNA. Therefore, a single-strand oligonucleotide of DNA primer hybridized to RNA can be extended until the end of the RNA is reached. Primer extension utilizes a primer of specific sequence to determine the amount and length of a specific RNA. This sort of analysis has been widely used to characterize mRNA made both in vivo and in vitro.

Primer extension is used to map the 5' termini of an RNA transcript; hence, the cap site of transcription can be determined. In addition, precursors and processing intermediates of mRNA can be detected. A short known fragment (30-80 bp) of DNA corresponding to a sequence near to the anticipated cap site of the gene of interest is labeled at the 5' terminus; this is termed the primer (a synthetic oligonucleotide or restriction fragment). The test RNA, which has been digested with RNase-free DNase to remove template DNA, can be synthesized in an in vitro transcription system (with the omission of the radiolabeled triphosphate) and hybridized with an excess of the single-stranded DNA primer. The primer and the RNA are then incubated at 85°C for 10 min to denature the nucleic acids, and annealed by incubating at a suitable temperature, for example, 55°C for 8-12 h. The exact tem-perature needs to be optimized by preliminary experiments. The length and GC content of the primer are used to determine the temperature of annealing; the standard temperature of 55°C is suitable for a 20 base oligonucleotide sequence with a GC content of approx 50%. The primer: RNA hybrid is then dissolved in reverse transcriptase buffer, approx 50 U of reverse transcriptase added to extend the primer to produce a cDNA copy complementary to the RNA template and incu-bated at 42°C for 90 min. If analyzing more than 10 μg of RNA in order to prevent distortion of the sequencing gel, a RNase digestion is needed at this stage.When the reverse transcriptase reaches the end of the sequence (the cap site), it "falls off." The length of the resulting end-labeled cDNA is measured by denaturing gel electrophoresis usually on a DNA sequencing gel, hence, the sequence prior to the cap site can also be determined.

Single-stranded DNA primers are preferred for experiments, since they eliminate the formation of DNA:DNA hybrids and can be designed to hybridize to specific sequences of the mRNA.

NUCLEASE S1 RNA MAPPING

The enzyme S1 nuclease degrades single-stranded DNA to yield 5' phosphoryl mono or oligonucleotide. Primer extension cannot be used to identify the 3' end of a transcript. Nuclease S 1 mapping is therefore the method of choice for this purpose. The procedure was originally described by Berk and Sharp in 1977; it has been used to map the location of 5' and 3' termini of mRNA or DNA templates, and locate 5' and 3' splice points.

End-labeled RNA or DNA probes derived from various segments of the genomic DNA are hybridized to mRNA favoring DNA: RNA hybrid formation. Products of the hybridization are digested with nucle-ase S 1 under conditions favoring single-stranded nucleic acid degra-dation only. Then the products are analyzed by gel electrophoresis to yield important quantitative and qualitative information about the mRNA structure. The basic method is as follows: Probe and RNA are mixed together, resuspended in hybridization buffer (40 mM PIPES [pH 6.4], 1 mM EDTA [pH 8.0],0.4M NaCl and 50% formamide), and incubated at 85°C for 10 min to denature the nucleic acids. Hybridiza-tion follows for 12-16 h, but again the temperature is dependent on the GC content of the DNA. Nuclease S 1 at 100-1000 U/mL and single-stranded carrier DNA at 20 pg/mL are then added for single-strand digestion, but the temperature depends on the degree of digestion required. Loops of DNA will be degraded at 20°C, and 37-45°C will digest single-stranded regions of DNA:RNA hybrids. The reactions are cooled to 0°C and nuclease stop mixture (containing carrier RNA at 50 $\mu g/cm^3$) added. The products can then be phenol: chloroform extracted, and the nuclease S1-resistant hybrids analyzed by alkaline or neutral agarose gel electrophoresis. In the case where a radiolabeled DNA probe has been chosen, this can be analyzed by electrophoresis through a polyacrylamide/urea gel and an autoradiogram established. A detailed description of nuclease-S 1 mapping was presented in a review by Favalaro et al. in 1980.

CELL AND NUCLEAR EXTRACTS

Soluble Cell Extract Transcription System for RNA Polymerase II

A soluble transcription system is a cell extract that will faithfully transcribe an exogenous DNA as template. For this, the extract should contain all the necessary factors needed for the correct initiation and elongation of the RNA chain, and be low in nuclease activity to pre-vent degradation of the exogenous DNA template or transcription products during the incubation. Establishing soluble systems provides a starting point for the separation, isolation, and analysis of those factors required (in addition to RNA polymerase) for accurate tran-scription in vitro.

To date, many eukaryotic and viral genes have been sequenced, and putative controlling sequences have been deduced. These observa-tions can be verified by in vitro manipulation of the DNA template and then using the DNA in an in vitro transcription system to test the effect.

Weil et al. in 1979 prepared a high-speed supernatant (S 100) cellular extract from cultured human cells. This S 100 extract can be used for transcription experiments once supplemented with purified

RNA polymerasc II. RNA polymerase II from wheat germ does not function in this system, indicating that it is a combined effect of RNA polymerase II and specific factors that mediate accurate and selective transcription. Fractionated S 100 does contain multiple factors required for accurate transcription by purified RNA polymerase II.

HeLa Whole Cell Lysate

The extract is prepared by gentle lysis of cells with a Dounce homog-enizer. Ammonium sulfate is added to approx 20% saturation, and the nuclei lysed. Cell debris and broken nuclei are removed by centrifu-gation at 175,000g for 3 h. Proteins in the supernatant are resuspended and dialyzed for 12 h. The dialysate is centrifuged to remove insoluble material, and supernatant removed to form the extract. This can be stored in small aliquots with repeated thawing and refreezing for up to 1 yr at—70°C. The extract normally contains 6—30 mg/mL of protein.

The rate of RNA polymerase II elongation is approx 300 nucleotides/min approx 1/10th that in vivo. The in vitro temperature of 30°C should not be increased to 37°C, otherwise RNase degradation may become a problem.

Each whole cell extract must be characterized by DNA and extract titration, since there is a threshold DNA concentration below which no accurate transcription occurs and an inhibitory effect at high DNA concentrations. Specific transcription can be obtained when the final concentration of protein in the extract is in the range 6-18 mg/mL. The extract contains high levels of 18S and 28S rRNA. Therefore, following extraction, it is possible that only 50-75% of the sample may be loaded onto the gel to avoid overloading. Since the extract is made from whole cells, weak promoters may not be as easily detected, although this does depend on the level of background nonspecific transcription occurring.

Nuclear Extracts from Human Cells

The extract is prepared by isolating nuclei in hypotonic buffer, lysis of cells with a Dounce homogenizer, extraction of the nuclei at 0.42M NaCl, high-speed centrifugation to clear the supernatant, and concen-tration of the extracted nucleic acids and various transcription factors by ammonium sulfate precipitation. Since the nuclei are isolated at very low ionic strength (0.0 1M KC1), retention of nuclear components is facilitated and contaminating cytoplasmic proteins are discarded in the supernatant following pelleting of the nuclei. Extraction at 0.42M

NaCl appears to release all of the nuclear components required for specific transcription, including RNA polymerase II, without releasing tightly bound nuclear proteins that can inhibit specific transcription.

Nuclear Extract

This method uses lysolecithin (a natural membrane lipid-lysophos-phatidylcholine) to disrupt plasma membranes and requires no deter-gents or lysis with a Dounce homogenizer. Lysolecithin nuclear extracts are said to be competent for RNA polymerase II and III transcription, DNA replication, pre-mRNA splicing, and sequence-specific DNA-protein binding. Nuclear extracts can be prepared on a small scale (10^7 cells), as well as for preparative purposes by this method.

Lysolecithin is highly toxic and cells lyse within 90 s. An extract made from HeLa cells at a density of 4-6 $\times$ 10^5/mL provides a suitable nuclear extract for in vitro transcription analysis. Using HeLa suspen-sions at 10^8 cells/mL, 100% lysis can be obtained with a concentration of lysolecithin at 300-800 μg/mL. In the protocol, a concentration of 400 μg/mL was chosen.

12

Reverse Transcriptase

The conversion of mRNA into cDNA is the essential first step in the study of eukaryotic cell products expressed from cloned genes. The key enzyme used first in this process, retroviral RNA-directed DNA polymerase (reverse transcriptase), catalyzes the synthesis of a DNA copy of an RNA template in the presence of a suitable primer. Reverse transcriptase (RT) was discovered in 1970 and was first used to copy a eukaryotic cell mRNA in 1971; the first cDNA clones prepared using RT were reported in 1976. Until recently, the only enzyme available was purified from avian myeloblastosis virus (AMV). The overall quality and consistency of commercially available AMV RT preparations have improved dramatically in the last six years, although considerable differences in performance characteristics still exist among enzyme preparations from different commercial suppliers. In addition, parameters for carrying out first-strand cDNA synthesis from poly(A)$^+$ mRNA populations with AMV RT have been optimized thoroughly; standard optima exist for pH, nucleotide concentration, and monovalent and divalent cation concentration; also essential stimulatory additives (such as sodium pyrophosphate and spermidine-HCl) have been identified. Furthermore, conditions have been established for proper handling and storage of the enzyme. Conditions for second-strand cDNA synthesis with first-strand product synthesized by AMV RT have also been optimized.

In 1985, Moloney murine leukemia virus (M-MLV) RT was cloned, overexpressed, and purified from *E. coli.* A series of reports have been published delineating optimal conditions for use of this alternative enzyme in first- and second-strand cDNA synthesis. M-

MLV RT has been used widely to synthesize cDNA and is frequently used instead of AMV RT for carrying out cDNA synthesis.

In 1989, a modified version of M-MLV RT lacking RNase H activity became available. RT without interfering RNase H activity catalyzes more efficient synthesis of full-length cDNA than either native AMV or cloned M-MLV RT possessing RNase H activity. The enzyme lacking RNase H activity is designated M-MLV H- RT. We describe here some of the properties of both forms of M-MLV RT and protocols for using M-MLV H- RT in first- and second-strand eDNA synthesis.

THE ENZYME

Sources

Commercial preparations of purified M-MLV RT are derived from two sources. Both proteins differ from the native enzyme in the amino acids at their amino and carboxy termini because of changes in sequence information introduced into the gene during the cloning procedure. The protein encoded by plasmid pB6B 15.23 has 25 additional amino acids fused at the amino terminus, is missing seven amino acids present at the carboxy terminus of the native enzyme, and has nine additional carboxy-terminal amino acids not found in the M-MLV *pol* gene. The RT encoded by pRT601 has one additional amino acid (methionine) at its amino terminus, the 671 amino acids present in the native enzyme,

Table 12.1 : Comparison of Some Properties of the Commercially Available RNase H-containing M-MLV RTs.

Plasmid encoding enzyme	Enzyme specific activity, U°/mgb	Units used typically in cDNA synthesis	Molar ratio of enzyme to mRNAC
pRT601	350,000	200	5.1
pB6B 15 23	35,000	20	5 1

and 12 additional amino acids at its carboxy terminus. The same number of molecules of each enzyme is needed to catalyze cDNA synthesis, despite their structural and apparent catalytic distinctions. Ten times as many units of pRT601 RT as pB6B 15.23 RT are required to copy efficiently an mRNA population. Not surprisingly, the specific activity of pRT601 RT (350,000 U/mg of protein) is 10 times greater than that of pB6B 15.23 RT (35,000-40,000 U/mg of protein) in the same unit assay. Since commercial preparations of both proteins are nearly homogeneous, an equal number of molecules of each enzyme is used during cDNA synthesis.

M-MLV H⁻ RT is encoded by a modified p601 plasmid and includes the same first 497 amino acids in cloned M-MLV RT, plus five additional amino acids at the carboxy terminus encoded by the *pol* gene read out of frame. The specific activity of M-MLV H^- RT is 100,000 U/mg of protein.

Unit Assay

The same unit assay is used by all manufacturers to define the DNA polymerase activity of RT. Poly(A) • oligo$(dT)_{12-18}$ is used as the template-primer, and the unit is defined as nmol of dTMP incorporated in 10 min at 37°C. The unit assay reaction conditions are not optimal for copying mRNA with either AMV, M-MLV RT, or M-MLV H^- RT.

Storage

Cloned M-MLV RT and M-MLV H^- RT are stable for long periods of time (>6 mo) when stored at -20°C in a buffer containing 0.1M NaCl, nonionic detergent, and 50% glycerol. Freezing the enzyme at -70°C does not affect the unit activity, but does decrease the functional activity (ability to copy long mRNAs). When stored in 50% glycerolcontaining solutions, the enzymes should never be placed at temperatures at which 50% glycerol freezes.

Catalytic Properties

There are two major difficulties with RT-catalyzed cDNA synthesis: (1) mass yields of cDNA transcripts generally do not exceed 50%, and (2) many cDNA transcripts are less than full length. When AMV RT was first used to synthesize cDNA, these difficulties were caused in part by RNase contaminants in AMV RT preparations. However, commercial preparations of AMV RT are now much cleaner. The high level of expression of cloned M-MLV RT and M-MLV H^- RT in E. *coli* has made it possible to eliminate detectable RNase contamination from enzyme preparations. However, results still fall short of the ultimate objective of synthesizing one full-length cDNA copy of every mRNA in a population, and this is owing in part to the catalytic properties of all RTs.

DNA Polymerase

RT resembles other template-directed DNA polymerases. It catalyzes the synthesis of DNA in the presence of a preformed primer with a 3'-OH hydrogen bonded to a template polynucleotide, a divalent metal ion, and a mixture of all four dNTPs. In contrast to most other DNA polymerases, RT can use either RNA or DNA to prime DNA synthesis from an RNA or DNA template. As a DNA polymerase,

RT displays low processivity and elongates DNA chains slowly. RT tends to pause at certain sequence and secondary structure-specific domains in RNA. As discussed below, such pausing can prevent an mRNA molecule from being copied all the way to its 5' end.

There is no endogenous 3'-5' exonuclease editing function associated with RT. In spite of this, the error rate of cloned M-MLV RT is low (1 mismatched residue in 30,000 incorporated). When the RNase H activity of cloned M-MLV RT was deleted to produce M-MLV H$^-$ RT, the error rate did not change.

RNase H

The RNase H activity of native RT specifically degrades the RNA strand of RNA • DNA and can act independently of the RT DNA polymerase activity. The initiation of cDNA synthesis depends on the hybridization of a DNA primer to mRNA, usually at the poly(A) tail. When this occurs, the hybrid is not only a priming site for the polymerase activity of RT, but is also a substrate for the RT RNase H activity as well. This sets up a competition between the two activities, and the extent to which the RNase H activity destroys the hybrid prior to the initiation of polymerization determines the maximum number of priming events that can actually occur. This reduces the yield of cDNA by effectively removing a portion of the mRNA from the reaction.

When RT is synthesizing cDNA, RT RNase H activity degrades the template that has already been copied because it is in hybrid form. If the scissions in the mRNA occur near the point of chain growth, the uncopied portion of the mRNA can dissociate from the transcriptional complex, resulting in termination of cDNA synthesis for that template. This problem is exacerbated by the tendency of RT to pause during reverse transcription and results in truncated cDNA molecules.

When M-MLV H$^-$ RT is used, RNA cleavage and its deleterious effects are eliminated, resulting in more efficient synthesis of fulllength cDNA. However, M-MLV H$^-$ RT still tends to pause during polymerization. Thus, some RNA molecules are still not copied to their 5' ends by M-MLV H$^-$ RT, even though the RNA remains intact during DNA synthesis. Other modifications to the RT polypeptide or addition of accessory proteins (e.g., a nucleic acid binding protein) will be required to reduce pausing and termination of transcription.

RNA Template

The quality of the mRNA template in a cDNA synthesis reaction dictates the maximum amount of sequence information that can be converted into cDNA. Thus, it is important to optimize the isolation of mRNA from a given biological source and to prevent adventitious introduction of RNases into a preparation that has been carefully rendered RNase-free. We recommend procedures that rely on chaotropic salts to inactivate RNases for extraction of RNA from any source. The presence of the poly(A) tail at the 3' end of mRNA has been exploited to fractionate mRNA from a total RNA population by affinity chromatography on oligo(dT) cellulose. Most cDNA libraries are made from poly(A)-selected mRNA. When the quantity of the RNA source is limiting and affinity chromatography is not feasible, it is possible to synthesize effectively cDNA from unfractionated mRNA. This is done routinely in preparation for amplifying cDNA by the polymerase chain reaction technique. Both M-MLV RT and M-MLV H^- RT are much less sensitive than AMV RT to inhibition by tRNA and rRNA, and well suited to copying unfractionated mRNA.

PRIMER

The primer for cDNA synthesis is usually oligo$(dT)_{12\text{-}18}$ present in molar excess relative to the 3' poly(A) tails of mRNA. This method of priming usually results in a cDNA library in which the 3' ends of RNAs are overrepresented because of the catalytic properties of RT. This bias can be partially overcome by using random hexamer primers instead of oligo(dT) or as a supplement to an oligo(dT)primed reaction. When random hexamers are used, the priming events are more evenly distributed over entire mRNA molecules. However, some sequence information corresponding to the 3' end of mRNA is lost as the result of randomizing primer location. In addition, rRNA in an RNA population is also copied when random hexamers are used.

The ratio of random hexamer primers to mRNA is critical. Figure 1 illustrates that there must be a balance between the desired average length and the mass yield of cDNA product. With M-MLV H^- RT, approx 10 primers/mRNA gives a reasonable yield without sacrificing product length. The ratio is higher with M-MLV RT, approx 40:1. When oligo(dT) or random primers are used to prime first-strand cDNA synthesis, the termini of the ensuing double-stranded cDNA are functionally identical, precluding directional cloning of the cDNA. All members of a directional library contain cDNA inserts cloned in a specific orientation relative to the transcriptional polarity of the original mRNA. Directional cloning is desirable when cDNA libraries will be screened by expression and is essential in some strategies for generating subtracted, i.e., partial, cDNA libraries.

cDNA can be cloned directionally by introducing two different restriction endonuclease recognition sites at its termini. This is most simply achieved by using a primer-adapter to initiate first-strand synthesis. A primer-adapter has the structure 5'-p$(dX)_n$-$(dT)_{15}$-OH-3', where $(dX)_n$ encodes sequence information for one or more restriction endonuclease sites. The primer-adapter introduces desired restriction endonuclease site(s) at the end of double-stranded cDNA corresponding to the 3' end of mRNA. Subsequent ligation of linkers or adapters to the DNA can be used to introduce a different restriction endonuclease site(s) at the opposite end of the DNA.

Reaction Conditions

pH and Metal Ion Requirements

Table elsewhere in this chapter gives the monovalent and divalent metal ion requirements of M-MLV RT and M-MLV H^- RT. The Mg^{2+} optima are quite sharp at 3 mM. KCI concentrations between 0 and 100 mM give product yields that are 75% or greater of those achieved at the optimum, 75 mM. Although the pH activity curve is broad, the yield of full-length product from a long RNA template (7.5 kb) for both enzymes is maximal at pH 8.3 at 22°C (at 37°C, pH 8.0).

Table 12.2 : Optima of M-MLV RT and M-MLV H^- RT.

$MgCl_2$, mM	*% Activity H^+*	*% Activity H^-*	*KCl, mM*	*% Activity H^+*	*% Activity H^-*	*pH, 22°C*	*% Activity H^+*	*% Activity H*
1	68	49	0	75	93	7.2	90	ND
2	88	93	50	95	100	7.5	95	ND
3	100	100	75	100	98	7.9	97	ND
4	88	ND	100	86	83	8.3	100	ND
5	83	77	125	72	66	8 5	98	ND
7.5	62	ND	150	54	44			
10	57	58	200	23	6			
15	ND	33						
20	20	15						

dNTPs

The K_ms of M-MLV RT for dCTP and dATP are 31 and 24 μM, respectively. The K_ms of M-MLV H^- RT for dNTPs have not been determined, but are assumed to be similar. In spite of these relatively low values, high concentrations (≥ 500 mM) of dNTPs in a reaction mixture result in the most efficient synthesis of full-length cDNAs.

Sodium Pyrophosphate

Sodium pyrophosphate at a concentration of 4 mM stimulates the synthesis of full-length cDNA from long mRNAs by AMV RT, although there is disagreement on the mechanism by which sodium pyrophosphate produces this effect. Sodium pyrophosphate also inhibits hairpin primed double-stranded cDNAsynthesis by AMV. Sodium pyrophosphate inhibits the polymerase activity of M-MLV RTs and should not be added to reactions catalyzed by these enzymes.

Polyanions

Spermidine-HCI at 0.5 mMhas been reported to stimulate the activity of AMV RT and Rauscher murine leukemia virus RT. However, spermine and spermidine inhibit both M-MLV RTs.

Bovine Serum Albumin and RNase Inhibitors

In many cDNA synthesis protocols, it is recommended that bovine serum albumin be added to reaction mixtures to stabilize RT. Bovine serum albumin has no effect on the activity of the M-MLV RTs. This may be because of the high level of RT protein added to a typical cDNA synthesis reaction.

The cloned M-MLV RTs are purified free of detectable RNase contamination. If precautions are taken to avoid adventitious introduction of RNase contaminants into reaction mixtures, RNase inhibitors are not necessary, and addition of such inhibitors can increase the chances of introducing contaminants into a reaction mixture.

Actinomycin D

Actinomycin D at 50-100 μg/mL inhibits the synthesis of doublestranded from single-stranded DNA. It is added during first-strand cDNA synthesis to inhibit hairpin primed double-stranded DNA synthesis. It probably interacts with 3' single-stranded cDNA ends in such a way that DNA • DNA hybridization needed for hairpin formation is inhibited. In first-strand cDNA reactions catalyzed by M-MLV RT, the proportion of cDNA molecules generated with exposed 3' singlestranded ends can be controlled to a large extent in the absence of actinomycin D by adjusting the ratio of enzyme to mRNA in the reaction. No actinomycin D is required in reactions catalyzed by M-MLV H$^-$ RT, since first-strand cDNA product remains hybridized to intact mRNA template, and hairpin primed DNA synthesis cannot occur.

Inhibitors

Glycerol concentrations as high as 19% can be tolerated in RT reaction mixtures. Both M-MLV RTs are inhibited by phosphate, pyrophosphate, and polyamines.

Reducing Agent

Both M-MLV RTs require dithiothreitol (DTT) for optimal activity. A minimum concentration of 5 mM is necessary, although 10 mM are optimal. Aqueous, unbuffered solutions of 0.1M DTT are stable at 20°C, even after repeated freeze-thaw cycles. However, DTT is oxidized quite rapidly in buffered solutions above pH 7.5, so that reaction buffers that include DTT should be prepared freshly and discarded after use. No data are available on the effect of P-mercaptoethanol on M-MLV RTs.

Temperature

When either M-MLV RT or M-MLV H^- RT is used to synthesize first-strand cDNA, the cDNA libraries with the largest average insert size result when first-strand cDNA synthesis is performed at 37°C.

Table 12.3 : Comparison of the Reaction Optima of M-MLV RT, M-MLV H^- RT, and AMV RT.

	Enzyme		
Parameter	***M-MLV RT***	***M-MLV H^- RT***	***AMV RT***
pH		8.3	8.3 8.4
[KCl]	75 mM	75 mM	50-100 mM
[$MgCl_2$]	3 mM	3 mM	10 mM
Temperature	37°C	37°C (42°C)°	42°C
[DTT]	10 mm	10 mm	1 mM
[dNTPs]	0.5 mM	0 5 mM	1 mM
Other components	None	None	4 mM NaPPi 0.5 mM Spermidine-HCl

M-MLV RT and M-MLV H^- RT can be used at incubation temperatures as high as 45 and 50°C, respectively.

THE SECOND-STRAND REACTION

We have chosen to describe exclusively a protocol for a one-tube format double-stranded cDNA synthesis procedure because it is fast, maximizes DNA product recovery, and yields product that can be modified and cloned into most vectors. The secondstrand cDNA is synthesized by nick-translation replacement of mRNA. First described by Okayama and Berg, and later popularized by Gubler and Hoffman, second-strand synthesis is catalyzed by *E. coli* DNA polymerase I in combination with *E. coli* RNase H and *E. coli* DNA ligase. Inclusion of E. *coli* DNA ligase in the reaction has been shown to improve the cloning efficiency of double-stranded cDNA synthesized from longer (>2 kb) mRNAs.

We must make several additional comments concerning the secondstrand reaction. The reaction is incubated at 16°C to reduce the tendency of DNA polymerase I at higher temperatures to strand displace rather than nick translate. The pH of the first-strand reaction is reduced in the second-strand reaction from 8.3 to 7.5 to suppress the 5'-3' exonuclease activity of DNApolymerase I. At higher pH, the 5'-3' exonuclease will attack exposed 5' single-stranded DNA ends more readily; in particular, the 5' end of a primer-adapter used to prime first-strand cDNA synthesis can be degraded extensively enough so that the restriction endonuclease recognition sites are lost. This can lower the output of clones if the cloning strategy relies on cleavage of one of these sites. Furthermore, the amount of sequence information corresponding to the length of the 5'-most RNA primer in the reaction normally lost during synthesis will presumably be minimized by the decrease to pH 7.5. The minimum number of bases lost because of the combined action of RNase H and DNA polymerase I 5'-3' and 3'-5' exonuclease is approx 8. Finally, T4 DNApolymerase is added at the end of second-strand synthesis to ensure that the termini of the double-stranded cDNA are blunt.

EXPERIMENTAL PROCEDURES

The following sections refer to conditions optimized for M-MLV H^- RT.

Laboratory Practices

It cannot be overemphasized that successful cDNA synthesis demands an RNase-free environment at all times. In general, this will require the same level of care used to maintain aseptic conditions when working with microorganisms. In addition, there are several guidelines that should be followed:

1. Never assume that anything is RNase-free except for sterile pipets, centrifuge tubes, culture tubes, and similar labware that is explicitly stated to be sterile.

2. Avoid using any recycled bottles and other glassware unless they have been specifically rendered RNase-free by rinsing with 0.5N NaOH followed by copious amounts of sterile, distilled water.

Alternatively, glassware can be baked at 150°C for 4 h.

3. Dedicate laboratory glassware for use with RNA and clearly mark it. Do not send anything outside of the laboratory to be washed once it has been rendered RNase free.

4. Microcentrifuge tubes can generally be taken from an unopened box, autoclaved, and used for all cDNA work. If necessary, they can be soaked overnight in a 0.01 % aqueous solution of diethylpyrocarbonate (DEPC), rinsed with autoclaved, distilled water, and then autoclaved.

5. If made with RNase-free labware, most solutions can be made from reagent-grade materials and distilled water, and autoclaved. Solutions that are heat-sensitive should be made with autoclaved, distilled water and sterile-filtered to 0.2 pm in disposable plasticware.

6. If all else fails, most aqueous buffer solutions can be treated with 0.01 % (v/v) DEPC and autoclaved. Buffers containing primary amines (such as Tris) cannot be effectively treated by this method.

7. Dedicate a separate set of automatic pipets for manipulating RNA and the buffers and enzymes used to synthesize cDNA.

8. Always wear disposable gloves to prevent contamination.

Materials

Reagents, Supplies, and Equipment

1. M-MLV H^- RT(BRL) (SuperScript™ RT).
2. *E. coli* DNA polymerase I.
3. *E. coli* DNA ligase.
4. *E. coli* RNase H.
5. T4 DNA polymerase.
6. $pd(N)_6$ (Random Hexamers)
7. $pd(T)_{12\text{-}18}$ [oligo$(dT)_{12\text{-}18}$]
8. [α-^{32}P]dCTP (>3000 Ci/mmol).
9. GF/C glass fiber filters (1 × 2 cm).
10. dATP, dCTP, dGTP, dTTP.
11. Autoclaved 0.5- and 1.5-mL microcentrifuge tubes.
12. Automatic pipets capable of dispensing 1-20 μL and 20-200 μL.
13. Autoclaved, disposable tips for automatic pipets.
14. Disposable gloves.
15. 37 and 16°C water baths.

16. Phenol:$CHCl_3$:isoamyl alcohol (25:24:1 [v/v/v]): Saturate a bottle of redistilled phenol with distilled water. When the phases have separated, remove some of the bottom layer (phenol) with a glass pipet and transfer it to a clean glass bottle. Add an equal vol of $CHCl_3$:isoamyl alcohol (24:1 [v/v]), and store at 4°C until use. This solution should not be kept for more than 1 wk. The water-saturated phenol should have the excess water removed from it before it is stored at -20°C until needed again.
17. 10% (w/v) TCA containing 1 % (w/v) sodium pyrophosphate (store at 4°C).
18 Yeast tRNA, 1 mg/mL: Place 25 mg of yeast tRNA (BRL) in 2.5 mL of *0.4M* NaCI-0.05M Tris-HCl (pH 7.5); add 0.1 mL of 10% sodium dodecyl sulfate and 1 mL of 1 mg/mL protemase K (BRL). Incubate at 37°C for 2 h. Extract at room temperature three times with an equal vol of phenol:$CHCl_3$:isoamyl alcohol, and precipitate the tRNA from the aqueous phase with 2.5 vol of ethanol. Dissolve the tRNA pellet in 0.4M NaCI-0.05M Tris-HCI, and repeat the precipitation. Dissolve the pellet in sufficient autoclaved, distilled, deionized water to bring thc concentration to 1 mg/mL (17.7 A_{260} U/mg). Store the solution at -20°C.
19. Autoclaved, distilled water.
20. Ethanol: 95% ethanol.
21. Bovine serum albumin (BSA), 50 mg/mL.

Solutions

1. dNTPs (10 mM): Prepare 50-mM stocks of each dNTP by dissolving 13-15 mg of each sodium salt in 100 μL of H_2O. Adjust to pH 6-7 by adding the appropriate amount of 1N NaOH (20-40 μL). Use pH paper. Add 4 μL of 1M Tris-HCI (pH 7.5) to buffer the solution, and bring the volume of the solution to 500 μL with water. Check the concentration of each solution at a 1/1000 dilution in water by determining the absorbance at 260 nm (E_{260} [1 cm/pH 7] for dCTP, dGTP, dATP, and dTTP is 7.4, 12.8, 15.3, and 8.4 cm^{-1} mM^{-1}, respectively). Prepare a mixture of all four dNTPs in water at a final concentration of 10 mM of each dNTP. Store all solutions at -20°C
2. 7.5 M Ammonium acetate.

3. 0.1 M Dithiothreitol (DTT) (store at -20°C).
4. 0.5 M EDTA: Adjust pH to 7.5 with NaOH.
5. 5X First-Strand Buffer: 250 mM Tris-HC1 (pH 8.3 at room temperature), 375 mM KCI, and 15 mM $MgC1_2$ (store at -20°C).
6. 5X Second-Strand Buffer: 94 mM Tris-HCI (pH 6.9 at room temperature), 453 mM KCI, 23 mM $MgC1_2$, 750 μM β-NAD, and 50 mM $(NH_4)_2\ SO_4$ (store at -20°C).
7. 10X DNA Synthesis Buffer: 200 mM Tris-HCI (pH 8.3 at room temperature), 500 mM KCI, 25 mM $MgC1_2$, and I mg/mL BSA (store at -20°C).

Model Protocols

One-Tube Double-Stranded cDNA Synthesis:

First-Strand Reaction

1. In a 1.5-mL tube, mix on ice 1-5 μg poly$(A)^+$ mRNA and one of the following:
   - 0.5-2.5 μg $pd(T)_{12-18}$
   - 40-200 ng $pd(N)_6$ or
   - A 10-fold molar excess of a primer-adapter (0.2-1.0 μg $pd[N]_{30}$) Add water to a vol of 7 μL; heat at 70°C for 10 min and chill on ice.
2. In the tube on ice, mix 4 μL 5X First-Strand Buffer, 2 μL 0.1M DTT, I μL 10 mM dNTPs, 1 μL $[\alpha\text{-}^{32}P]$dCTP (1 μCi/μL), and sufficient water to bring the final vol to 20 μL after RT is added, if second-strand cDNA is to be labeled, omit the $[\alpha\text{-}^{32}P]$dCTP.
3. If $pd(T)_{12-18}$ or primer-adapter is present, equilibrate the tube to 37°C before adding 200-1000 U (1-5 μL) of M-MLV H^- RT (SuperScript™ RT, use 200 U/μg mRNA); if $pd(N)_6$ is used, immediately add the appropriate amount of RT, and incubate at room temperature for 10 min; finally incubate all reaction types at 37°C for 1 h.
4. The final reaction composition is 50 mM Tris-HCI (pH 8.3),75 mM KC1, *3* mM $MgC1_2$, 10 mM DTT, 500 pM each of dATP, dCTP, dGTP, and dTTP, 50-250 μg/mL mRNA, either 25-125 μg/mL $pd(T)_{12-18}$, 210 μg/mL $pd(N)_6$, or 10-50 μg/mL 30 base primer-adapter, and 10,000-50,000 U/mL SuperScript™ RT.
5. Place the tube on ice. Remove 2 μL from the reaction mixture,

and add them to 48 μL of 20 mM EDTA (pH 7.5) containing 5 pg of tRNA. This will be used to calculate first-strand yield and to analyze the product by gel electrophoresis.

6. Use the remaining 18 μL of first-strand reaction mixture (or the entire 20-μL reaction mixture if ^{32}P was not used) to carry out second-strand synthesis.

One-Tube Double-Stranded cDNA Synthesis: Second-Strand Reaction

1. This protocol is suitable for 1-5 pg mRNA originally in the 20-μL firststrand reaction mixture.
2. On ice, add the following reagents in the order shown to the first-strand reaction mixture tube: water to bring the final reaction mixture vol to 150 μL, 30 μL 5X Second-Strand Buffer, 3 μL 10 mM dNTPs, 10 U *E. coli* DNA ligase, 40 U *E. coh* DNA polymerase I, and 2 U E. *colt* RNase H. If the second-strand product is to be labeled, add 10 μCi of [α^{32}P]dCTP after the unlabeled dNTPs. Vortex the tube gently, and incubate at 16°C for 2 h.
3. The final composition is 25 mM Tris-HCl (pH 7.5), 100 mM KCI, 5 mM $MgC1_2$, 10 mM $(NH_4)_2SO_4$, 0.15 mM (3-NAD, 250 pM each of dATP, dCTP, dGTP, dTTP, 1.2 mM DTT, 67 U/mL DNA ligase, 267 U/mL DNA polymerase I, and 13 U/mL RNase H.
4. Add 20 U of T4 DNA polymerase, and continue incubating at 16°C for 5 min.
5. Place the tube on ice. If [α-^{32}P]dCTP was added to the second-strand reaction, remove 10 pL from the tube and add it to 40 μL of 20 mM EDTA containing 5 pg of tRNA. This will be used to calculate yield and to analyze the products by gel electrophoresis.
6. Add 10 pL of 0.5M EDTA to the remaining reaction mixture on ice. The product can now be phenol-extracted and ethanol-precipitated in preparation for addition of linkers or adapters.

One-Tube Double-Stranded cDNA Synthesis: First-Strand Reaction

1. The following protocol is designed for synthesis of first-strand cDNA suitable for PCR amplification with specific primers. For strategies designed for PCR amplification of total cDNA libraries.

2. In a 0.5-mL tube, mix on ice 1 μg total RNA (smaller amounts of poly [A]$^+$ mRNA can be used) and one of the following: 0.5 μg pd(T)$_{12\text{-}18}$, or 100 ng μd(N)$_6$. Add water to a vol of 14 μL; heat at 70°C for 10 min and chill on ice.
3. In the tube on ice, mix 2 μL 10X DNA Synthesis Buffer, 2 μL 0.1M DTT, and 1 μL 10 mM dNTPs.
4. If pd(T)$_{12\text{–}18}$ is used, equilibrate the tube to 37°C, and add 200 U (1 μL) of SuperScript™ RT; if pd(N)$_6$ is used, immediately add 1 μL of SuperScnpt™ RT, and incubate at room temperature for 10 min; finally, incubate all reaction types at 37°C for 1 h.
5. The final reaction composition is 20 mM Tris-HC1 (pH 8.3), 50 mM KCI, 2.5 mM MgC1$_2$, 10 mM DTT, 500 pM each of dATP, dCTP, dGTP, and dTTP, 50 pg/mL RNA, 25 μg/mL pd(T)$_{12\text{–}18}$ or 5 μg/mL pd(N)$_6$, 100 pg/mL BSA, and 10,000 U/mL SuperScript™ RT.
6. Heat the tube at 90°C for 5 min, place the tube on ice for 5 min, and collect the tube contents by brief centrifugation.
7. Add 2 μL of RNase H (2 U/μL) to the tube, and incubate for 20 min at 37°C before proceeding with PCR amplification. Digestion of the firststrand cDNA-RNA hybrid to remove RNA is essential to subsequent amplification of the cDNA.
8. Place the tube on ice. Add directly to the tube, and mix 8 pL I0X DNA Synthesis Buffer, appropriate amounts of PCR primers, sufficient water to bring the final vol to 100 μL after *Taq* DNA polymerase is added, and *Taq* DNA polymerase (1-4 U).
9. Layer 100 μL of mineral oil over the reaction mixture.
10. Carry out PCR amplification.

[^{32}P]cDNA Synthesis

1. The following protocol is designed to synthesize high specific activity (4 × 10^8 cpm/gg) [^{32}P]cDNA that represents all sequences in an RNA population. The ratio of random primer to RNA template has been selected to maximize product yield and length. Such labeled cDNA can be used as a hybridization probe or as starting material for generating a subtracted cDNA library.
2. In a 1.5-mL tube, dry 10 μL [α-^{32}P]dCTP (10 μCi/pL, >3000 Ci/mmol). Add on ice 4 μL 5X First-Strand Buffer, 2 μL

0.1M DTT, 1 μL 10 mM each of dATP, dGTP, and dTTP, 4 μL 0.1 mM dCTP, 0.5 jig mRNA, 0.3 pg pd(N)$_6$, and sufficient water to bring the final vol to 20 μL after RT is added. Gently vortex the tube, and add 0.5 pL (100 U) of M-MLV H$^-$ RT. Incubate at room temperature for 10 min and at 37°C for I h.

3. Add 5 μL 0.5M EDTA and an equal vol (25 μL) of 0.6N NaOH, and incubate at 68°C for 30 min. Remove unincorporated dNTPs by chromatography in 0.1M NaCl, 10 *mM* Tns-HC1(pH 7.5), and 0.1 *mM* EDTA over a Sephadex G-50 column. Recover the product by ethanol precipitation.
4. Approximately 70% of the total dCTP in the reaction should be incorporated, resulting in the synthesis of 1.5 × 10^1 cpm of labeled product from 0.5 pg of mRNA. Up to 5 pg of RNA can be copied in a 20-μL reaction. If more than 0.5 pg of RNA is used, the amount of dCTP, [α^{32}P]dCTP, pd(N)$_6$, and RT used should be increased proportionately

Product Analysis

First-Strand Reaction

1. Spot duplicate aliquots (10 μL) from the diluted sample on separate glass fiber filters. Dry one filter and count in scintillant to determine the specific activity of the dCTP in the reaction mixture. Wash the other filter in TCA-sodium pyrophosphate, and dry and count the filter to determine yield. The yield of the first-strand reaction is calculated from the amount of acid-precipitable radioactivity determined. In order to perform the yield calculation, the specific activity of the [α-^{32}P]dCTP in the reaction must be determined. The specific a^:ivity (SA) is defined as the counts per minute (cpm) of an aiiquot of the reaction mixture divided by the quantity (pmol) of the same nucleotide in the aliquot.

$$\text{SA (cpm/pmol dCTP)} = (\text{cpm}/10\ \mu\text{L})/(200\ \text{pmol dCTP}/10\ \mu\text{L}) \quad (1)$$

The SA should be approx 200 cpm/pmol. Once the SA is known, the amount of cDNA in the first-strand reaction can be calculated from the amount of acid-precipitable radioactivity determined from the washed filter:

$$\text{Amount of cDNA } (\mu\text{g}) = [(\text{cpm}) \times (50\ \mu\text{L}/\ 10\ L) \times (20\ \mu\text{L}/2\ \mu\text{L}) \times (4\ \text{pmol dNTP/pmol dCTP})]/[(\text{cpm/pmol dCTP}) \times (3030\ \text{pmol dNTP}/\mu\text{g cDNA})] \quad (2)$$

The correction in the numerator takes into account that, on the average, four nucleotides will be incorporated into the cDNA for every dCTP. The factor in the denominator is the amount of nucleotide that corresponds to I μg of single-stranded DNA.

2. Precipitate the product in the remaining diluted sample by adding 0.5 vol of 7.5M ammonium acetate and 2 vol of cold ethanol, and then centrifuging. Wash the pellet with 70% ethanol, centrifuge again, and dry the pellet after removing the ethanol. The size of the cDNA product can be analyzed by alkaline agarose gel electrophoresis. Dissolve the sample pellet in 10 μL sample buffer (30 mM NaOH, 1 mM EDTA, 10% glycerol, 0.01 % bromophenol blue). Appropriate [^{32}P] DNA size markers should be placed in sample buffer, e.g., BRL 1-kb DNA Ladder. The gel (1.4% [w/v]) should be cast in 30 mM NaCl, 2 mM EDTA, and then equilibrated for at least 3 h in alkaline electrophoresis buffer (30 mM NaOH, 2 mM EDTA) before loading the samples. For an 11 × 14 cm horizontal gel, electrophoresis should be for 5-6 h at 50 V or for 16-18 h at 15 V. Dry the gel under vacuum, heat for 1-2 h, and expose the gel to X-ray film overnight at room temperature.

Second-Strand Reaction

1. Spot duplicate aliquots (10 μL) from the diluted sample on separate glass fiber filters, and proceed with determining the [α-^{32}P]dCTP specific activity and the amount of acid-precipitable radioactivity. The SA of [α-^{32}P]dCTP is given by:

$$\text{SA (cpm/pmol dCTP)} = (\text{cpm}/10\ \mu\text{L})/(500\ \text{pmol dCTP}/10\ \mu\text{L}) \quad (3)$$

The specific activity should be approx 500 cpm/pmol. The amount of cDNA synthesized in the second-strand reaction is given by:

$$\text{Amount of cDNA } (\mu\text{g}) = [(\text{cpm}) \times (50\ \mu\text{L}/10\ \mu\text{L}) \times (150\ \mu\text{L}/10\ \mu\text{L}) \times (4\ \text{pmol dNTP/pmol dCTP})]/[(\text{cpm/pmol dCTP}) \times (3030\ \text{pmol dNTP}/\mu\text{g cDNA})] \quad (3)$$

2. Ethanol precipitate the DNA from the remaining diluted sample and proceed with alkaline agarose gel.

INDEX

B

D

F

G

H

N

O

P

Q

U

V

W

X

Y

Z